Internet of Drones

This book covers different aspects of Internet of Drones (IoD) including fundamentals in drone design, deployment challenges, and development of applications. It starts with a detailed description of concepts and processes in designing an efficient system, and architecture. It details different applications of IoD and its implementations in smart cities, agriculture, health care, defense, security, logistics, GIS mapping, and so forth. Recent developments in IoD design, application of AI techniques, case studies, and future directions are covered.

Features:

- Focuses on important perspectives of the Internet of Drones (IoD)
- Emphasizes drone deployment in smart cities, smart agriculture, smart health care, and 3D mapping
- Covers challenges in drone design for applications with security and privacy issues
- Reviews diversified drone applications with real-use cases from modern drone players ranging from start-up companies to big giants in the drone industry
- Includes different aspects of drone design such as hardware and software architecture, potential applications, and opportunities

This book is aimed at researchers and professionals in computer sciences, electronics and communication engineering, and aeronautical engineering.

Internet of Drones
Applications, Opportunities, and Challenges

Edited by

Saravanan Krishnan and M. Murugappan

CRC Press
Taylor & Francis Group
Boca Raton London New York

CRC Press is an imprint of the
Taylor & Francis Group, an **informa** business

Designed cover image: © Shutterstock

First edition published 2023
by CRC Press
6000 Broken Sound Parkway NW, Suite 300, Boca Raton, FL 33487-2742

and by CRC Press
4 Park Square, Milton Park, Abingdon, Oxon, OX14 4RN

CRC Press is an imprint of Taylor & Francis Group, LLC

ISBN: 978-1-032-17168-5 (hbk)
ISBN: 978-1-032-17169-2 (pbk)
ISBN: 978-1-003-25208-5 (ebk)

DOI: 10.1201/9781003252085

Typeset in Times
by MPS Limited, Dehradun

Contents

About the Editors

Dr. Saravanan Krishnan is working as a senior assistant professor in the Department of Computer Science and Engineering at Anna University, Regional Campus, Tirunelveli, Tamil Nadu. He has received an M.E. in software engineering and PhD in computer science engineering. His research interests include cloud computing, software engineering, Internet of Things, and smart cities. He has published papers in 14 international conferences and 27 international journals. He has also written 14 book chapters and edited eight books with international publishers. He has done many consultancy works for municipal corporations and smart city schemes. He is an active researcher and academician. He is a member of ISTE, IEI, ISCA, ACM, etc.

He has also trained and interviewed the engineering college students for placement training and counseling. He has conducted many ISTE workshops in association with IIT Bombay and IIT Kharagpur. He is also coordinator for IIRS (Indian Institute of Remote Sensing) outreach programme. He has delivered more than 50 guest lecturers in many seminar/conference in reputed engineering colleges.

Professor M. Murugappan has been working at Kuwait College of Science and Technology (KCST), Kuwait as a full professor in electronics, Department of Electronics and Communication Engineering, since 2016. He also serves as a visiting professor at the School of Engineering at Vels Institute of Science, Technology, and Advanced Studies in India. He is also an international visiting fellow at the Center of Excellence in Unmanned Aerial Systems at Universiti Malaysia Perlis in Malaysia. In 2006, he graduated from Anna University, India with an ME degree in applied electronics. He received his PhD from Universiti Malaysia Perlis, Malaysia in 2010 for his contribution to the field of mechatronic engineering. Between 2010 and 2016, he worked as a senior lecturer at the School of Mechatronics Engineering, Universiti Malaysia Perlis, Malaysia. In this role, he taught a variety of courses related to biomedical and mechatronics engineering. For his excellent publications and research products, he has been a recipient of several research awards, medals, and certificates. In a study by Stanford University, he was recently ranked among the top 2 percent of scientists working in experimental psychology and artificial intelligence for three consecutive years (2020–2022). His research into affective computing has received more than 750K in grants from Malaysia, Kuwait, and the UK. His publications include more than 140 peer-reviewed conference proceedings papers, journal articles, and book chapters. Several of his journal articles have been recognized as best papers, best papers of the fiscal year, etc. Professor Murugappan is a member of the editorial boards of *PLOS ONE, Human Centric Information Sciences, Journal of Medical Imaging and Health Informatics*, and *International Journal of Cognitive Informatics*. In addition to being the Chair of the IEEE Kuwait Section's Educational Activities Committee, he is an active reviewer for IEEE Transactions on Multimedia, *IEEE Transactions on Affective Computing, IEEE Transactions on Health Informatics*, and *IEEE Transactions on Biomedical Signal Processing*. He is interested in affective computing, affective neuroscience, the Internet of Things (IoT), the Internet of

Medical Things, cognitive neuroscience, brain–computer interface, neuromarketing, neuroeconomics, medical image processing, machine learning, and artificial intelligence. Throughout his career, he had a strong collaborative relationship with leading researchers in the fields of biosignal/image processing, artificial intelligence, the IoT, and cognitive neuroscience. He is a member of several international professional societies including IEEE, IET, IACSIT, IAENG, and IEI.

Contributors

C. Aasish
Department of Aeronautical Engineering
KCG College of Technology
Chennai, Tamil Nadu, India

Jose Anand
Department of Electroncis and Communication Engineering
KCG College of Technology
Chennai, Tamil Nadu, India

K. Anirudh
Dayananda Sagar University
Bangalore, Karnataka, India

T. Arulmozhinathan
Department of Aeronautical Engineering
Hindusthan College of Engineering and Technology
Coimbatore, Tamil Nadu, India

R. Asad Ahmed
Department of Aeronautical Engineering
KCG College of Technology
Chennai, Tamil Nadu, India

J.V. Bibal Benifa
Department of Computer Science and Engineering
Indian Institute of Information Technology
Kottayam, Kerala, India

R. Dinesh
SNS College of Technology
Coimbatore, Tamil Nadu, India

R. Ganesan
School of Computer Science and Engineering
Vellore Institute of Technology
Chennai, Tamilnadu, India

S.K. Geetha
Department of Computer Science and Engineering
KPR Institute of Engineering and Technology
Coimbatore, Tamil Nadu, India

V.T. Gopinathan
Department of Aeronautical Engineering
Hindusthan College of Engineering and Technology
Coimbatore, Tamil Nadu, India

Akash Ilangovan
Madras Institute of Technology
Anna University
Chennai, India

M.K. Karthik
Department of Aeronautical Engineering
Bharath Institute of Higher Education and Research
Chennai, India

P. Karthikeyan
Department of Production Technology
Madras Institute of Technology
Chennai, India

R. Maheswari
School of Computer Science and Engineering
Vellore Institute of Technology
Chennai, Tamilnadu, India

S. Mano
School of Mechanical Engineering
SASTRA Deemed to be University
Thanjavur, India

C. Muralidharan
Department of Computing Technologies
SRM Institute of Science and Technology
Kattankulathur, India

V. Nagaraj
Periyar Maniammai Institute of Science and Technology
Thanjavur, India

S. Syam Narayanan
Department of Aeronautical Engineering
KCG College of Technology
Chennai, Tamil Nadu, India

R. Naveen
Bannari Amman Institute of Technology
Erode, India

P. Pandiyan
Department of Electrical and Electronics Engineering
KPR Institute of Engineering and Technology
Coimbatore, Tamil Nadu, India

Varalakshmi Perumal
Madras Institute of Technology
Anna University
Chennai, India

G. Prasad
Chandigarh University
India

S. Praveena
Department of Aeronautical Engineering
Bharath Institute of Higher Education and Research
Chennai, India

Kavitha Rajamohan
School of Sciences
CHRIST (Deemed to be University)
Bangalore-29, Karnataka, India

Sakthi Jaya Sundar Rajasekar
Melmaruvathur Adhiparasakthi Institute of Medical Sciences and Research
Melmaruvathur, India

Y. Rajkumar
School of Information Technology and Engineering
Vellore Institute of Technology
Vellore, India

J. Bruce Ralphin Rose
Department of Mechanical Engineering
Anna University Regional Campus
Tirunelveli, Tamil Nadu, India

K. Saravanan
Department of Computer Science and Engineering
Anna University Regional Campus
Tirunelveli, India

D.D. Shivaprasad
International University of East Africa
Kampala, Uganda

V.M. Sreehari
School of Mechanical Engineering
SASTRA Deemed to be University
Thanjavur, India

N. Srirangarajalu
Department of Production Technology
Madras Institute of Technology
Chennai, India

M. Subramanian
Department of Science and Humanities
KPR Institute of Engineering and Technology
Coimbatore, Tamil Nadu, India

Rajasekaran Thangaraj
Department of Computer Science and Engineering
KPR Institute of Engineering and Technology
Coimbatore, Tamil Nadu, India

Kanagaraj Venusamy
Department of Mechatronics
Rajalakshmi Engineering College
Thandalam, Chennai, India

S. Vivekanandan
Department of Electrical and Electronics Engineering
KPR Institute of Engineering and Technology
Coimbatore, Tamil Nadu, India

Mohamed Sirajudeen Yoosuf
School of Computer Science and Engineering
VIT-AP University
Amaravati, Andra Pradesh, India

Preface

In recent years, the Internet of Things (IoT) has now emerged as a promising technology in a variety of applications. IoT can be used in flying drones to detect, track, and deliver goods and environment surveillance. IoT empowers drones as intelligent, thin devices by utilizing cognitive computing techniques. IoT makes drones a light-weighted flying device by minimizing the use of heavy electrical and mechanical components. A drone connected with IoT resulted in the "Internet of Drones (IoD)." Several potential applications have been developed by researchers in the design of drones using IoT in the literature. Specifically, IoD plays a vital role in the defence department, healthcare monitoring and emergency applications, delivery of goods, sports photography, object movement detection, crowd monitoring and analysis, traffic surveillance, agriculture surveillance, etc. The purpose of this edited book is to cover three important perspectives of the Internet of Drones (IoD): fundamentals in drone design, challenges in IoD deployment, and opportunities in developing real-time applications. This book covers a 360-degree spectrum of different verticals in which drone applications are gaining momentum. The wide variety of topics it presents would offer the readers a good understanding of IoD. The different chapters in the edited book will cover the above three important perspectives in IoD.

1 Internet of Drones: Applications, Challenges, Opportunities

J. Bruce Ralphin Rose, T. Arulmozhinathan, V.T. Gopinathan, and J.V. Bibal Benifa

CONTENTS

DOI: 10.1201/9781003252085-1

1.1 INTRODUCTION

UAVs are otherwise known as unmanned aerial vehicles in cases if operated in the atmosphere or autonomous underwater vehicles (AUVs) if operated in the underwater environment without any pilot onboard. As the unmanned vehicles are operated remotely using electronic controls, UAVs are implemented in areas where the human workforce will be facing risks or difficulty in access. UAVs are said to have been used from the mid-1800s for warfare, and the earlier UAVs were developed as balloons, torpedoes, and aerial targets at those times. Since then, militaries around the world have been equipped with UAV technology for training, target practice, airstrikes, bomb detection, surveillance, and hostage negotiation. The earliest recorded UAV attack was held during July 1849 in Venice by the Austrian navy using 200 incendiary balloons. Since then, the world has been continuously making efforts to develop the UAV systems for warfare and related applications.

Only in 2006 were UAVs first deployed for non-military purposes. Government agencies quickly began to evaluate the UAV technologies for disaster relief and border surveillance activities while corporations began to use them for commercial applications like pipeline inspections, crop quality evaluation, and security. Apart from the hobby UAVs, the first legal delivery of commercial supply was recorded in the year of 2015. The UAV market is expected to reach USD$40.7 million by 2026, according to the present trends [1].

On the other hand, the evolution of sensors and actuators through Industry 4.0 is rapid and enormous. The integration of sensors and actuators with day-to-day objects via a communication platform or Internet is known as Internet of Things (IoT) and adds a new level to modern technology. The influence of IoT is revolutionizing every aspect of drone technology through compatible solutions. With immense improvement in material science, integrated circuits, and communication aerodynamics, UAVs are much more affordable to be implemented in several applications that support commercial business as well apart from defense alone. The autonomy of UAVs in their application is certainly soon and the bioinspired designs can deliver more promising value-added solutions. The working nature of the UAV requires strong communication and exchange of data between the controller and itself, thus requiring a cloud/data access platform for its operation. Adding to it, to operate a UAV on its own it is necessary that the controller has to make judgments based on the information communicated from the UAV and communicate back to the UAV for its operation [2].

1.1.1 How IoDs Work?

UAVs have multiple applications in defense as well as in civilian airspace. In defense, UAVs are used for surveillance, monitoring a critical area, locating a specified target, rescue, medical supply, disaster management (i.e., natural disasters), as well as wartime spoils, and attacking and defending targets. Likewise, UAVs are being implemented in a huge list of civilian applications such as product

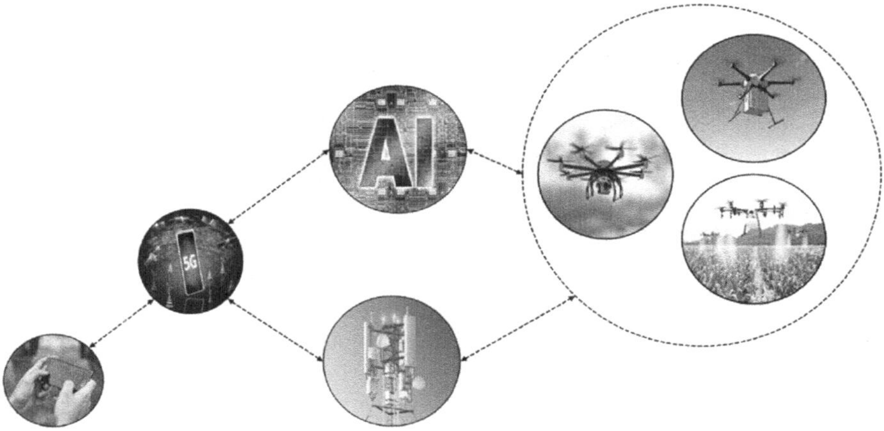

FIGURE 1.1 Internet of Drones and their component systems.

delivery, firefighting, smart agriculture, photography in the small time as well as film making, assisting police in surveillance, emergency medical supply, and many more [3]. Particularly, the response time would be reduced by about 50% with the use of drones during disaster management and recovery activities.

It is good to mention a simple example on how these applications would be integrated through the Internet at this point. In case of a fire breakout on a building, the surveillance UAV, surveillance camera, mobile SOS app would communicate the firefighting station through the cloud or dedicated communication layer and trigger the firefighting UAV for immediate assistance. The firefighting UAV will be supported by the input sensors that will identify the type, hotspots, and intensity of the fire and make judgments to put it out. At the same time, it will also send the information on the current situation to the human firefighters who approach the incident.

In any such stated application, a UAV requires a large volume of data for its operation alone; alongside it collects a separate set of data based on its application. The Internet on the other hand has the capability to handle large volumes of data to apply to things physically, which is known as Internet of Things (IoT) [4]. Thus, it is inevitable to connect the UAVs to their application through the Internet and typical IoD functions and their component systems are displayed in Figure 1.1.

1.2 APPLICATIONS OF IODS

In every application of drones, there is a lot of data that will be gathered and communicated during their operation to complete the mission. Based on the application, the volume and nature of data collected would be different and it requires unique methods for classification and processing. At any instance of the application, there may be a single UAV or multiple UAVs that are operated to accomplish the mission [5]. Some of the crucial applications of IoDs are listed as follows.

1.2.1 Warfare

The main aspect for the UAV is to reduce human fatality during defense operations. The modern advanced UAV is capable of monitoring the critical zones like national borders, forest areas, and different terrains. It can attack the targets and also defend them by detecting and destroying the bogeys that trespass the borders.

1.2.2 Disaster Management

One of the important roles of a UAV during disaster management is the emergency response and damage assessment to mitigate the losses. UAVs are used to identify and locate the extent of disaster and to deliver emergency supplies to the people affected by the disaster. A rescue UAV locates a stranded target at an unknown location with appropriate rescue tool kits. A swarm of autonomous drones will be more suitable for such purposes.

1.2.3 Smart Surveillance

Smart surveillance is an IoD-oriented essential application where a UAV is deployed to monitor a human-congested festival area or a mining site or a construction site. Surveillance on coastlines, ports, bridges, and other access points for the import of illegal substances is the remit of the customs and excise authority. Patrolling the fishing area in the sea to alert the civil fisherman about abnormal weather status can also be done through IoD-enabled communication systems [6].

1.2.4 Smart Agriculture

Agriculture is one of the prime zones where the UAVs can be implemented as a modernized smart farming tool. Monitoring the crops, spraying pesticides and fertilizers, seed sowing, animal herding, and neutralizing human wildlife encounters are some fields where precision piloting must be done to assist the farmers. In such a case, a large volume of data must be acquired, and a decision must be made by the UAV [7]. The recently manufactured drones become smarter by integrating open-source technology and smart sensors and have more flight endurance to track down animals and crops [8,9].

1.2.5 Firefighting

One of the most dangerous risks involved tasks in firefighting is the identification of hotspots, which may end up in human casualties. A remote-controlled drone or an automated drone using a thermal imaging process can identify the intensity of the flame and counter it with proper fire extinguishing techniques. Even forest fires can be controlled under this technique through a swarm of drones equipped with fire extinguisher balls and chemical agents [4].

Many other applications like aerial photography, monitoring pollution from high altitude, inspection of power lines laid by electric companies for maintenance service, assisting local police bodies in crowd monitoring, survey organization in surveying landscapes, mapping large cities in a digital format, and helping to study the river course in case of floods from aerial point of view. Deep sea diving, and sewage rate monitoring can also be done using a UAV at the required localities to minimize the environmental hazards [6].

1.3 COMPONENTS OF IODS

The functionalities of IoDs as described in section 1.2 consist of the following sub-systems as listed in Table 1.1.

Each system is configured with its device with proper security and capable of functioning individually and integrate to support the other systems. The IoD applications and deliverables are illustrated in Figure 1.2.

1.3.1 UAV Parts in Construction

The fundamental UAV components and details one should be familiar with are as follows.

TABLE 1.1
The Functionalities and Sub-Systems of IoDs

Sl.No.	Sub-Systems	Purposes
1.	Sensing systems	It identifies each UAV from the other in the IoD system
		Senses the environment using appropriate sensors and camera
2.	Communication systems	Defines the communication in the IoD system
3.	Computation systems	Responsible for processing and computational ability of the IoD system
4.	Services systems	Represents all categories of services provided by the IoD system

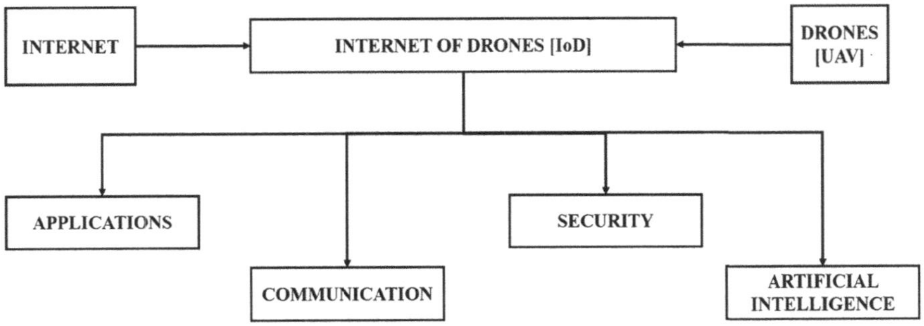

FIGURE 1.2 Internet of drones and their deliverables.

1.3.1.1 UAV Frames

It is the structure that acts as a skeleton where all other parts will be fitted. Based on the geometry, the UAV can be a fixed plane or a copter model. In both models, the parts are assembled uniformly to maintain the centre of gravity. They are available in various shapes, sizes, and price ranges depending on their application and quality. It also has a landing gear to land the UAV and park it [10].

1.3.1.2 Propellers

Propellers are rotating blades that produce a pressure difference and push the air from one side to the other side by creating a thrust force. The suction side of the rotor has a low-pressure envelope that initiates the thrust in the same direction. The lift generation is dependent to the position and pitch of the rotor. In the case of the fixed-wing UAV, the rotor is positioned at the nose section of the UAV as a single-engine configuration. Thus, it produces the required thrust and the wing produces lift in a similar manner to the conventional propeller-driven aircrafts, whereas in the copter model, the propellers are positioned in the vertical axis thus the lift is produced directly without a fixed wing. The number of propellers in a copter is decided by its application and weight to meet the lift requirements. Often the propellers are twin blade rotors made up of plastic or carbon fibres or glass fibres [11].

1.3.1.3 Motors

The propellers are rotated by either electric motors or tiny IC engines according to the UAV configuration and endurance requirements, whereas in case of some fixed-wing UAVs, miniature fuel engines are used like the trainer aircraft designs. It is quite efficient to connect each propeller to its own motor rather than coupling a single motor for multiple rotors. The motors are also fitted in such a way that the controller easily rotates them, as displayed in Figure 1.3. Their rotation enhances the UAV control in terms of direction. Choosing the right motor is essential for the efficiency of the UAV. Parameters such as voltage and current, thrust and thrust to weight ratio, power, efficiency, and speed must be checked carefully.

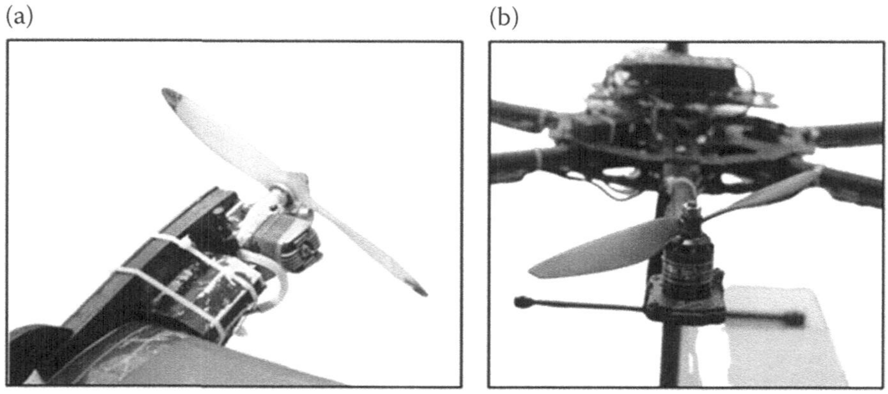

FIGURE 1.3 (a) Fixed-wing UAV with a fuel engine; (b) quadcopter UAV with its motor.

1.3.1.4 Electronic Speed Controller (ESC)

An ESC is an electronic control board that varies the rpm of the UAV propellers, and it also acts as a dynamic braking system. The component helps the ground pilot to approximate the height at which the UAV is running through the predefined thrust available-to-thrust required trends. This is attained by gauging the amount of power or thrust-to-weight (T/W) ratio used by all the motors. It should be noted that the altitude is associated with power drain from the power reservoirs [12].

1.3.1.5 Radio Transmitter

It is a channelled transmitter and a communicator to the UAV. Each channel has a specific frequency capable of steering the UAV in a certain motion. UAV requires at least four channels for effective operation.

1.3.1.6 Battery, Electronics, and Power Distribution Cables

The battery acts as the power source to the UAV and it supplies energy in all the electronics in the framework through the power distribution cables. Nickel metal hybrid or nickel cadmium batteries were first used; however, their use has diminished while the use of lithium batteries has been increased. They can store a greater amount of energy than the nickel cadmium and nickel metal hybrid counterparts. The rating of a typical battery is 3,000 mAh and 4 V. The electrical and electronic part is a crucial part pertaining to the control and operation of the UAV. However, in respect to the purpose of the UAV, other components can be either included or omitted. The UAV may be functional without these parts, though for multitasking purposes it's advised to include them [13].

1.3.1.7 Sensors and Camera

The UAV is attached with multiple sensors based on its application. In the case of a surveillance application, the UAV can be mounted with thermal imaging sensors and cameras for video footage. Cameras with the capability of shooting and storing or sending videos are available and used in relation to the operator's need.

1.3.1.8 Flight Control Board

The flight control board makes a log of the take-off place just in case the need arises for the UAV to go back to its take-off location without being guided. This is known as the "return to home" feature. It also determines and calculates the UAV's altitude in respect to the amount of power it consumes. UAVs need a controller to maintain stability and to reach their goal. The most well-known drone controllers are proportional-integral-derivative (PID) and proportional derivative (PD) controllers. However, the controller parameters need to be tuned and optimized. Modern evolutionary algorithms like biogeography-based optimization (BBO) and particle swarm optimization (PSO), for multi-objective optimization (MOO) can be used to tune the parameters of the PD controller of a drone. The combination of MOO, BBO, and PSO results in various methods for optimization such as vector-evaluated BBO and PSO, denoted as VEBBO and VEPSO, where the non-dominated sorting BBO and PSO are denoted as NSBBO and NSPSO. The multi-objective cost function is based on tracking errors for the four states of the system. However, NSBBO performs better than the other methods [14].

1.3.1.9 First-Person View (FPV)

The first-person view is often what the pilot is seeing through the UAV camera for controlling the UAV remotely. Though it is expensive in comparison to the normal control device interface (transmitter) screen, it gives the user an interactive 3D view experience. The first-person view (FPV) gives the user an ultimate feeling of as if the user is flying. There are two major components of an FPV system. The first one is the ground component. The ground component is also called the ground station. It consists of a video receiver and a display system on the ground. The second major component is the video receiver. The video receiver receives the data by matching the frequency of the receiver with the transmitter present of the UAV. The most common frequencies used for transmission of video are 433 MHz, 869 MHz, 900 MHz, 1.2 GHz, 2.4 GHz, and 5.8 GHz. Advanced versions of ground components have sophisticated antennas which result in greater image resolution. The airborne component has a camera and a video transmitter on the UAV. These are the basic components of a basic FPV system. Advanced FPV systems have advanced components and functions. For instance, the FPV system can be added with a GPS navigation system. It can also have flight data systems. Further advanced FPV systems have "return home" systems which allow the UAV to come back to the position from where they initially took off [10].

1.3.2 UAV Communication

Based on their application, a UAV can be operated as a single device or as a swarm of drones. In both the cases, the UAV must communicate to the ground station or cloud server and to any other UAV in the controlled airspace. Hence, the communication of a UAV is between themselves and to the ground station (Figure 1.4). A UAV-to-UAV communication is critical because of the frequency regulations

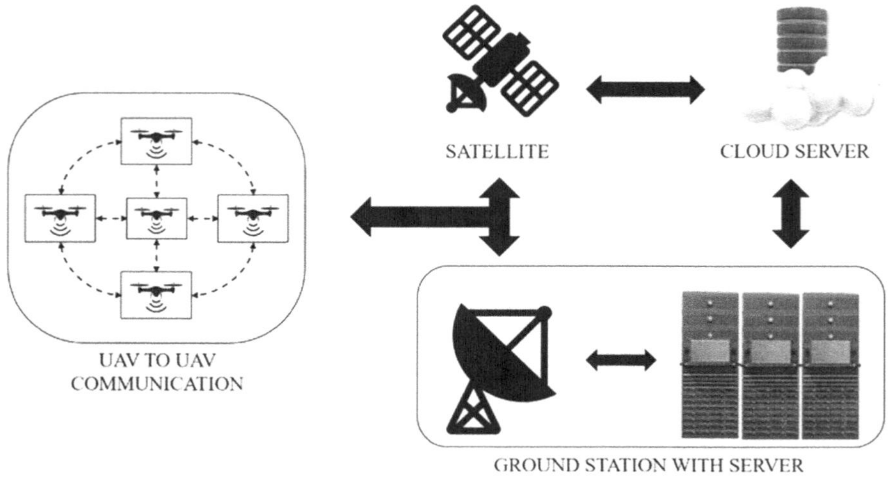

FIGURE 1.4 UAV communication method.

and airspace sharing. It especially important to communicate between the UAVs to avoid a collision and complete the mission when operating as a swarm.

1.3.2.1 UAV-to-UAV Communication

Airborne transponders are likely to be the best method to avoid collisions where the location coordinates and velocity information of the UAV are obtained and fed through an automatic dependent surveillance broadcast (ADS-B) sensing method. It helps to avoid collisions such as point of closest approach, Dubins paths approach, and collision cone approach [15]. Satcom link offers the UAV operators a throughput of 200 kbps for less than 2 kg weight out of which they can get improved aircraft control and situational awareness together with the possibility to perform real-time route adjustments. The main limitations are the size, weight, and power requirements for the UAV, thus putting this method to a confined boundary.

A Wi-Fi radio link can be modelled as the interconnectivity between ground control stations (GCS) and two UAVs. It can be implemented through a functional and reliable proof-of-concept to study the UAV communication relay system by using commercially available hardware (Wi-Fi) components combined with customized software. LTE data link is beneficial for low-altitude high interconnectivity. 4G/5G/LTE terrestrial networks are seen as the right and soonest available solution to interconnect UAVs and drone pilots while ensuring a correlated airspace sharing [16].

A UHF radio link can be proposed for UAV-to-UAV communication. Point-to-point wireless link in the UHF band (400 MHz) works reliably in near line of sight (nLOS) or in non-line of sight conditions (NLOS) while a Wi-Fi link in 2.4 GHz could suffer a drop of connectivity [17].

1.3.2.2 UAV to Ground Station

All UAVs must be controlled remotely from a ground station using wireless connectivity. They can be connected by dedicated radio signals, satellite signals or even by cellular base stations. Swarms of unmanned aerial vehicles (UAVs) can provide a plethora of services ranging from delivery of goods to surveillance. UAVs in a swarm utilize wireless communications to collect information, like velocity and heading angle, from surrounding UAVs for coordinating their operations and maintaining target speed and UAV-to-UAV distance. However, due to the uncertainty of the wireless channel, wireless communications among UAVs will experience a transmission delay that can impair the swarm's ability to stabilize system operation [18].

The swarm of UAVs in a mission must sense and transmit a large volume of data for their operation. Thus, a large volume of UAVs will require a much-intensified network with high-rate transmissions and the UAVs are expected to be a vital component of the upcoming 5G and beyond 5G (B5G) wireless networks that can potentially facilitate wireless broadcast and support high-rate transmissions. Compared to the communications with fixed infrastructure, a UAV has salient attributes, such as flexible deployment, strong line-of-sight (LoS) connection links, and additional design degrees of freedom with the controlled mobility. Subsequently, an exhaustive review of various 5G techniques can be provided based on UAV platforms in which it can be categorized by different domains including physical layer, network layer, and joint communication, computing, and caching [19].

1.3.2.3 IoT-Enabled UAV Communication System

Because of the limited processing capabilities and low on-board storage, drones are unable to perform computationally demanding applications. Integration of drones with IoT and the cloud is envisioned as a viable solution to this shortcoming. A service-oriented cloud-based management system, or drone planner, can use the micro air vehicle link (MAVLink) protocol for communication and provides a simple yet efficient application program interface to develop UAV applications. Alternatively, machine-type multicast service (MtMS) can be used for enabling the concurrent data transmission to machine-type communication devices. Its architecture and procedures are designed to optimize latency, reduce energy consumption, and control overhead. Trade-offs between turning agility, flying speed, and battery life have been analyzed with the help of these factors and various experiments. A buses-and-drones mobile infrastructure (AidLife) utilizes an existing public transport system to establish an adaptable system for reliable communication during a disaster. Additionally, physical collisions, IoT equipment selection, communication technology, efficient UAV networking, as well as regulatory concerns, has to be conferred. Moreover, cloudlets and computational offloading are one of the best solutions for efficient computing while conserving energy. Both IoT and body sensor Nntworks (BSNs) are among the main applications of 5G networks. Device discovery, which involves the registration, deletion, query, and route formation for nodes in a network can cause excessive energy consumption in nodes and hamper the operation of these networks. Extensible markup language (XML) energy charts have been employed to facilitate energy-efficient device discovery for 5G-based IoT and BSNs using UAVs [20].

1.3.3 Communication Requirements (Data)

UAVs are commonly employed in tasks where humans cannot be put to risk. Commonly, such areas are remote and have a barrier to communication. Thus, a strong communication is required to support the different UAV applications.

a. Widespread coverage from 10 m to 3,000 m altitude to monitor the UAV in the network for various application.
b. Real-time and remote communication to monitor the UAVs flight conditions, mission, equipment, and emergency control
c. Transmission of HD image/video will need a high Gbps level data rate. The 5G networks can provide support for transmitting high resolution images or 4K/8K HD video based on its application needs. With higher transmission rates, a UAV can provide a increased experience in the VR and AR domain.
d. Every UAV in the network has to be identified, monitored, forecast the flight situations, and coordinated [21].

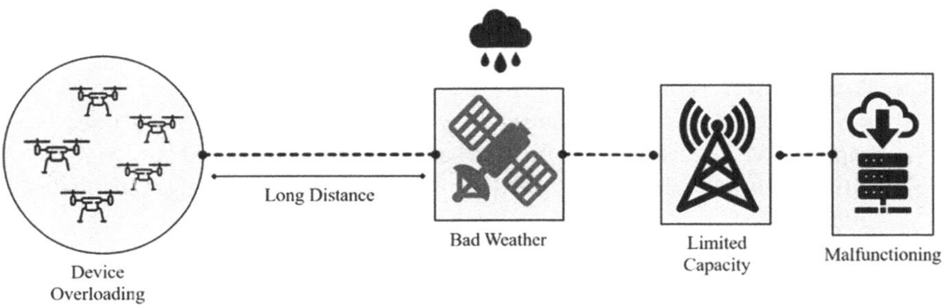

FIGURE 1.5 UAV communication latency.

Every UAV needs a strong communication without any latency issues. A UAV under various applications and missions has to minimize information inactivity that is the lag between the start of the mission and the instant when the data captured from the sensing location arrives to the base station and constrain latency that is the lag between capturing data at a sensing location and its arrival at the base station. This is important in many application scenarios where sensing locations should not only be visited as soon as possible, but the captured data needs to reach the base station in time, especially if the investigation area is larger than the communication range (Figure 1.5). In the case of multiple UAVs, the swarm cooperatively stores and transports the data forward along minimum latency paths (MLPs) to guarantee data delivery within a predefined latency bound. Also, MLPs specify a lower bound for any latency minimization problem where multiple mobile agents store and transport data [22].

With the advent of 5G network services, vehicular cloud networks that provide computation capabilities have taken precedence over traditional costly cloud solutions. However, with vehicular cloud computing, a variety of new challenges have been ascended. Recently, it was demonstrated that an intra-UAV resource sharing model could provide a range of cloud services for modern-day UAV applications [23].

1.3.4 Security Requirements

An IoD network must provide proper security and privacy for the UAV operations and data processing. The localization of a UAV is unreliable because of the privacy and security problems that come with the network. Thus, a critical objective of the IoD network is to enhance the security and privacy levels to a point where it cannot be negotiated. Consequently, the key necessities of an IoD network are legitimacy, privacy, accessibility, integrity, and non-repudiation.

a. **Authentication:** An authentication key is required before disclosing the essential information about the UAVs. The IoD network should use a network that produces the authentication key for every session using an incorruptible method.

b. **Confidentiality:** It is the process of making data available and controlling the access to that data safely and effectively by ensuring data privacy [24].
c. **Availability:** A registered UAV should be granted access with appropriate network services in an ongoing mission. An IoD network must be capable of governing the flight management system without compromising the availability criteria if an UAV is malfunctioning and distinguish it from other UAV in the network [24].
d. **Integrity and Non-Repudiation:** An IoD network must guarantee the trustworthiness of the information. It must be ensured that the networks activity is not concealed. All the activity must be monitored and cannot be rejected without proper conclusion [25].

All the devices and sensors that support the communication and operation of the drones are susceptible for getting hijacked or involved in human error, additionally the safety for autonomous UAVs is critical. Moreover, any kind of UAV implementation requires approval from the government authorities [26].

UAVs are aircrafts that are utilized to collect video recordings, still images, or live video of the targets, such as vehicles, people, or specific areas under the surveillance application. Habitually in battlefield surveillance, there is a high possibility of eavesdropping, inserting, modifying, or deleting the messages during communications among the deployed UAV and base station server (BSS). It leads to initiating several potential attacks by an adversary, such as main in-middle, impersonation, drones hijacking, replay attacks, etc. Moreover, anonymity and untraceably are the two crucial security properties that are needed to be maintained in the battlefield surveillance communication environment. Such a crucial security problem can be dealt with by an access control protocol for battlefield surveillance in IoD through the detailed security analysis. It is done by formal and informal (non-mathematical) security verification under automated software simulation tools and the testbed experiments for various cryptographic primitives. IoD encryptions can also resist several potential attacks advanced in battlefield surveillance scenarios [27].

UAVs are harmless and extremely useful when used properly; however, misappropriated UAVs lead directly to invasions of privacy, concerns with aircraft safety, and even result in a personal injury. An armed UAV used for military purposes could induce catastrophic results if it is being hacked. Small UAVs that are used by civilians are even easier to be attacked and hijacked. These UAVs use a wireless transmission protocol for navigation (radio waves or WiFi) and telemetry, and the navigation signals associated with this are easy to manipulate to hack the UAV. Many surveys show that the vulnerabilities including UAV hacks are based on unprotected WiFi, access to UAV configuration files and changes to settings in flight and GPS attacks.

WiFi insecurity: Some UAV applications allow multiple connections through WiFi, which allows an attacker to tamper with the system and hack the drone.

SkyJack: The SkyJack uses a malicious UAV to crack and enter into the other UAV's network using WiFi. The SkyJack disconnects the controller by changing the SSID and immediately reconnects and it is then able to transmit commands through the malicious UAV to the hijacked UAV by enabling it to take control over the entire UAV network in flight using a customized script.

GPS based attacks: It is well known that the UAVs use GPS signals for navigation. Typically, the GPS signals used by civilian UAVs are not encrypted and these UAVs are prone to GPS attacks frequently.

Maldrone: Maldrone sets up a proxy serial port that intercepts flight commands from the controller and redirects actual serial port communication to fake ports, while forwarding the hijackers commands to the drone [28].

Cooperative ad-hoc UAV networks need essential security services to ensure their communication security. Cryptography, as the inseparable tool for providing security services, requires a robust key management system. Unfortunately, the absence of infrastructure in cooperative networks leads to the infeasibility of providing conventional key management systems. Key pre-distribution schemes have shown promising performance in different cooperative networks due to their lightweight nature. However, the most concerning issues they suffer from are intermediate decryption encryption (DE) steps and the lack of key updates. They can utilize a highly dynamic UAV node movement in 3D space to provide the key update feature and optimize the number of intermediate DE steps [29].

As aggressive security initiatives, a UAV market focuses on frequency hopping, spectrum spreading, and key sharing. Legally, providers can only run-on industrial, scientific and medical (ISM) bands with the 2.4 GHz band that is dependent on packet-based transfer using networking approaches. A micro air vehicle link (MAVLink) is a wireless networking protocol that enables entities to communicate. It is used for bidirectional communications between the ground control station (GCS) and the UAVs. The UAV receives orders and controls from the GCS, while the GCS receives telemetry and status information from the UAV. Lorenz Meier first launched a MAVLink for real-time applications in early 2009 under the lesser general public license (LGPL). A single GCS can support up to 255 aircrafts using MAVLink. The MAVLink protocol's minimum packet length is 8 bytes, and the maximum packet length is 263 bytes with a complete payload. To transmit control and telemetry data, a bidirectional connection is needed. The aerial vehicle's telemetry data will be sent to the GCS, while control data will be sent the other direction.

Further, a byte-wise MAVLink message is transmitted over the contact channel, followed by an error-correcting checksum. If the checksums do not align, then the message is compromised and will be deleted. A packet start sign (STX) is used by MAVLink to sync the start of an encoded message as displayed in Figure 1.6. After receiving the packet start symbol, the packet length is read, and the checksum is checked after the total number of bytes. Subsequently, the decoded packet is processed with an acknowledgement message, and it waits for the next start sign to identify whether the checksum matches. A checksum loss caused by tampered or missing message bytes causes the packet to be discarded, and the receiving system resumes listening for the next start sign packet. As a protection measure for packet loss prevention, MAVLink assigns a sequence number (SEQ) to each packet. If the packet loss rate tends to be high, the pilot will order the UAV to return or at the very least limit its operational range, thus protecting the UAV from falling into the wrong hands [30].

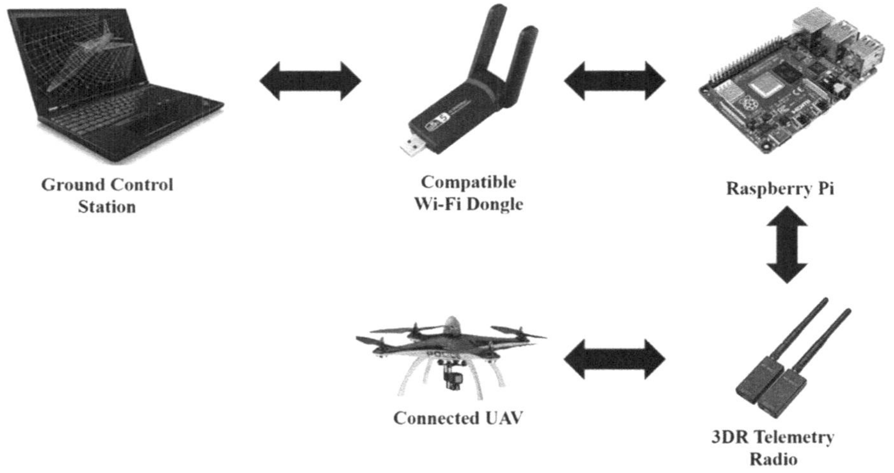

FIGURE 1.6 MAVLink network.

1.4 ARTIFICIAL INTELLIGENCE (AI) FOR IOD

An exceptionally large volume of data has to be received and computed for a smooth operation of UAV missions. Any latency or wrong decision will collapse the UAV swarm, eventually ending the purpose of the mission. To compute such large data, AI must be roped in. AI enables the machines to behave like humans through learning, recognizing patterns, decision making, and solving problems. It makes decisions by processing a large amount of complex organized and unorganized data obtained through a network of technologies and sensors, popularly known as the Internet of Drones (IoD) [31]. An IoD network helps in the communication and interaction of data with or without human interference. IoD produces a vast amount of data from a swarm of UAV and its sensors. Based on the application, the data may be collected not only from the surface-mounted sensors on the UAV but also from external sources like weather forecast systems, traffic monitoring cameras, surveillance cameras in the streets or buildings, and even from the satellite feeds. These data are analyzed in real time by employing various AI models to predict weather, calamities, accidents, agriculture, medical emergency, forest fire, building fires, and many more. AI models process and call decisions based on this large volume of data within an abbreviated period of time and transmits back to the UAV for smooth operation [32,33].

UAVs are widely used for forest fire monitoring and detection using learning algorithms. Machine learning (ML) algorithms such as decision trees (DTs), support vector machines (SVMs), and Naïve-Bayes (NB) classifiers are employed for fire detection by extracting key features. As compared to ML models, convolutional neural networks (CNN) combined with Yolo have offered a good recognition rate and confidence score as compared to traditional ML methods [34]. The high-level workflow of drone data capturing, subsequent data transfer to the cloud storage where it is preprocessed and fed to a deep learning (DL) model are highlighted in

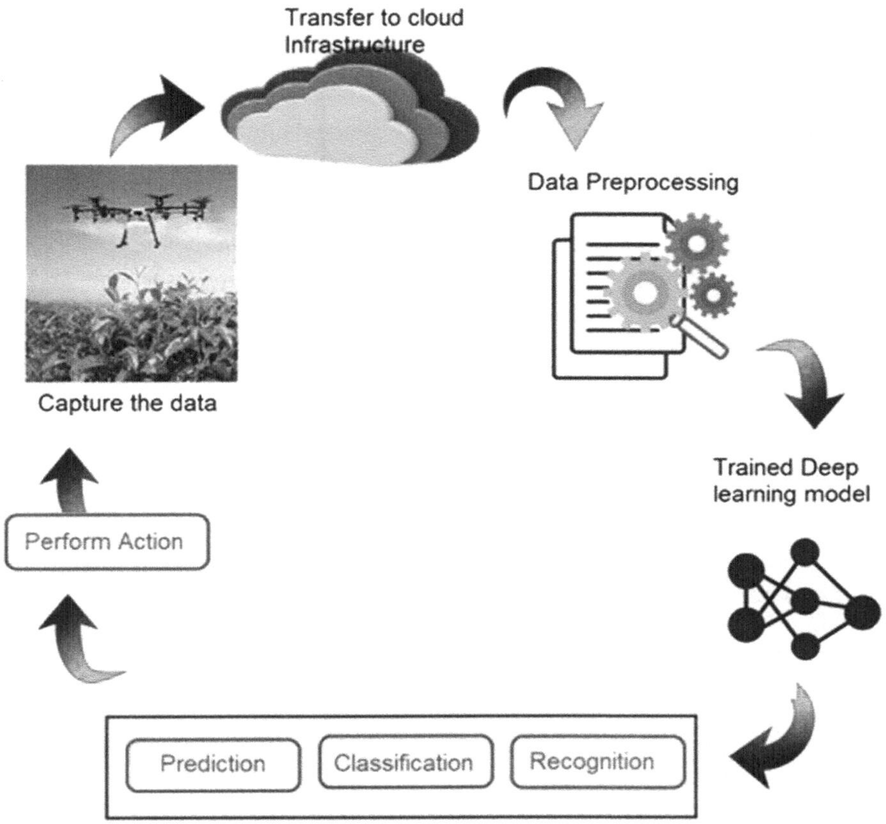

FIGURE 1.7 High-level workflow of drones using deep learning models.

Figure 1.7. Here, the DL model does the intended task such as detection, classification, recognition, or prediction [35]. Subsequently, the computed activity is accelerated by the drone during the operation.

Drones are used for pollination and spraying of pesticides so that there is a need for supervised learning models for the accurate identification of crops, disease affected areas, temperature monitoring, etc. Drones are also used for delivering the goods and parcels at various locations and there is an extensive need for identifying the buildings and road areas that can be done by deploying DL models using CNN. Drones are employed for aerial surveying and hence there is a need for classifying the different terrains such as lands, forests, and water bodies. DL models such as Yolo can be used to detect and classify objects in this regard. The advent of bioinspired designs will also enhance the real-time data acquisition possibilities at various complex environments where the human access is almost impractical [36]. Further, the bioinspired drones used for disaster management need learning algorithms for navigation, prediction, recognition, and classification. Hence, AI is a part of IoD where supervised learning algorithms is a mandatory component.

1.5 CONCLUSION

Unmanned aerial vehicles have many applications in the domestic as well as defense sectors and their implementation is rapidly growing throughout the world. The policymakers also relaxed several drone registration procedures to enhance the research activities in addition to the aerial photography, surveillance, geo-mapping survey, toys, and some of the customized industry applications like delivery, mining, and maintenance. With the advent of modern technologies like AI, cloud, and edge computing, 5G networks in Industrial Revolution 4.0 have been testifying the integration of various artefacts of the communication platform with the advancements of the sensors and actuators. Correspondingly, the substantial scientific developments in material science, integrated circuits, and aerodynamics have widely opened a paradigm where the Internet and drone technologies can be blended with a systematic architecture. A UAV commonly requires a remote-control operation by a human/robotic pilot, whereas in IoD, full autonomy of the UAV ensured with a swarm of drones can be achieved, which could entail a large volume of data for their successful operation. The communication security of such a large volume of data poses a challenge and it requires unique security patches to overcome the cyberattacks. However, by blending advanced communication security strategies with UAVs, IoD concepts enable more secure air space management for a wide range of applications. For instance, power consumption, augmented range, battery requirements, swarm functioning, lower altitude ATC, UAV framework, communication, and signal encryption could see a potential upgrade in the near future. Also, virtual reality shall be augmented to improvise the IOD technology and thus opening up a lot of opportunities based on applications and challenges in every domain where IoD is implemented.

REFERENCES

[1] https://percepto.co/the-evolution-of-UAVs-from-military-to-hobby-commercial/ - accessed on 15 Jan 2022.

[2] A. Kumar and B. Muhammad, 'On how Internet of drones is going to revolutionise the technology application and business paradigms', *Int. Symp. Wirel. Pers. Multimed. Commun. WPMC*, vol. 2018-November, pp. 405–410, 2018, doi: 10.1109/WPMC.2018.8713052.

[3] R.J. Hall, 'A geocast-based algorithm for a field common operating picture', *Proc. - IEEE Mil. Commun. Conf. MILCOM*, vol. 4, pp. 1–6, 2012, doi: 10.1109/MILCOM.2012.6415848.

[4] N. Jayapandian, 'Cloud Enabled Smart Firefighting Drone Using Internet of Things', *Proc. 2nd Int. Conf. Smart Syst. Inven. Technol. ICSSIT 2019, no. Icssit,* 2019, pp. 1079–1083, doi: 10.1109/ICSSIT46314.2019.8987873.

[5] M. Hassanalian and A. Abdelkefi, 'Classifications, applications, and design challenges of drones: A review', *Prog. Aerosp. Sci.*, vol. 91, no. April, pp. 99–131, 2017, doi: 10.1016/j.paerosci.2017.04.003.

[6] https://www.allerin.com/blog/10-stunning-applications-of-drone-technology - accessed on 07 Jan 2022.

[7] https://www.equinoxsdrones.com/blog/importance-of-drone-technology-in-indian-agriculture-farming - accessed on 04 Jan 2022.

[8] V. Puri, A. Nayyar, and L. Raja, 'Agriculture drones: A modern breakthrough in precision agriculture', *J. Stat. Manag. Syst.*, vol. 20, no. 4, pp. 507–518, 2017, DOI: 10.1080/09720510.2017.1395171.

[9] H. Hejazi, H. Rajab, T. Cinkler, and L. Lengyel, 'Survey of platforms for massive IoT', *2018 IEEE Int. Conf. Futur. IoT Technol. Futur. IoT 2018*, vol. 2018-January, pp. 1–8, 2018, doi: 10.1109/FIOT.2018.8325598.

[10] https://www.foundationstructures.com/10-UAV-parts-everybody-in-construction-should-know/ - accessed on 18 Jan 2022.

[11] G. Andria et al., 'Design and Performance Evaluation of Drone Propellers', *Proc. 5th IEEE Int. Workshop on Metrol. AeroSpace (MetroAeroSpace)*, 2018, pp. 407–412, doi: 10.1109/MetroAeroSpace.2018.8453604.

[12] https://www.dronezon.com/learn-about-drones-quadcopters/how-drone-motors-esc-propulsion-systems-work/ - accessed on 20 Jan 2022.

[13] M.N. Boukoberine, Z. Zhou, and M. Benbouzid, 'Power Supply Architectures for Drones - A Review', *Proc. 45th Annual Conf. IEEE Indus. Electron. Soc.*, 2019, pp. 5826–5831, doi: 10.1109/IECON.2019.8927702.

[14] A. Shamshirgaran, H. Javidi, and D. Simon, 'Evolutionary Algorithms for Multi-Objective optimization of Drone Controller Parameters', *Proc. IEEE Conf. Control Technol. Appl. (CCTA)*, 2021, pp. 1049–1055, doi: 10.1109/CCTA48906.2021.9658828.

[15] B. Li, Y. Jiang, J. Sun, L. Cai, and C.Y. Wen, 'Development and testing of a two-UAV communication relay system', *Sensors (Switzerland)*, vol. 16, no. 10, pp. 1–21, 2016, doi: 10.3390/s16101696.

[16] E. Kalantari, M.Z. Shakir, H. Yanikomeroglu, and A. Yongacoglu, 'Backhaul-aware robust 3D drone placement in 5G+ wireless networks', *Proc. IEEE Int. Conf. Commun. Work. ICC Work*, 2017, pp. 109–114, doi: 10.1109/ICCW.2017.7962642.

[17] G. Militaru, D. Popescu, and L. Ichim, 'UAV-to-UAV Communication Options for Civilian Applications', *Proc. 26th Telecommun. Forum, TELFOR*, 2018, pp. 1–4, doi: 10.1109/TELFOR.2018.8612108.

[18] T. Zeng, M. Mozaffari, O. Semiari, W. Saad, M. Bennis, and M. Debbah, 'Wireless communications and control for swarms of cellular-connected UAVs', *Conf. Rec. - Asilomar Conf. Signals, Syst. Comput.*, vol. 2018-Octob, pp. 719–723, 2019, doi: 10.1109/ACSSC.2018.8645472.

[19] B. Li, Z. Fei, and Y. Zhang, 'UAV communications for 5G and beyond: Recent advances and future trends', *IEEE Internet Things J.*, vol. 6, no. 2, pp. 2241–2263, 2019, doi: 10.1109/JIOT.2018.2887086.

[20] A. Sharma *et al.*, 'Communication and networking technologies for UAVs: A survey', *J. Netw. Comput. Appl.*, vol. 168, no. May, p. 102739, 2020, doi: 10.1016/j.jnca.2020.102739.

[21] G. Yang *et al.*, 'A Telecom Perspective on the Internet of Drones: From LTE-Advanced to 5G', pp. 1–8, 2018, [Online]. Available: http://arxiv.org/abs/1803.11048.

[22] J. Scherer and B. Rinner, 'Multi-UAV Surveillance With Minimum Information Idleness and Latency Constraints', vol. 5, no. 3, pp. 4812–4819, 2020, doi: 10.1109/LRA.2020.3003884.

[23] V. Balasubramanian, S. Otoum, M. Aloqaily, I. Al Ridhawi, and Y. Jararweh, 'Low-latency vehicular edge: A vehicular infrastructure model for 5G', *Simul. Model. Practice Theory*, vol. 98, p. 101968, 2020, ISSN 1569-190X, 10.1016/j.simpat.2019.101968.

[24] Y. Ko, J. Kim, D.G. Duguma, P.V. Astillo, I. You, and G. Pau, 'Drone secure communication protocol for future sensitive applications in military zone', *Sensors*, vol. 21, no. 6, pp. 1–25, 2021, doi: 10.3390/s21062057.

[25] M. Yahuza *et al.*, 'Internet of drones security and privacy issues: Taxonomy and open challenges', *IEEE Access*, vol. 9, pp. 57243–57270, 2021, doi: 10.1109/ACCESS.2021.3072030.
[26] J. Johnson, 'Artificial intelligence, drone swarming and escalation risks in future warfare', *RUSI J.*, vol. 165, no. 2, pp. 26–36, 2020, doi: 10.1080/03071847.2020.1752026.
[27] B. Bera, A.K. Das, S. Garg, M.J. Piran and M.S. Hossain, 'Access control protocol for battlefield surveillance in drone-assisted IoT environment', *IEEE Internet of Things Journal*, doi: 10.1109/JIOT.2020.3049003.
[28] V. Dey, V. Pudi, A. Chattopadhyay, and Y. Elovici, 'Security Vulnerabilities of Unmanned Aerial Vehicles and Countermeasures: An Experimental Study', pp. 0–5, 2018, doi: 10.1109/VLSID.2018.97.
[29] M. Algharib and F. Afghah, 'How UAVs Highly Dynamic 3D Movement Improves Network Security?' https://arxiv.org/abs/2105.02608, 2021.
[30] H.M. Ismael, Z. Tariq, and M. Al-ta, 'Authentication and encryption drone communication by using HIGHT lightweight algorithm', *Turkish Journal of Computer and Mathematics Education*, vol. 12, no. 11, pp. 5891–5908, 2021.
[31] Z. Lv, 'The security of Internet of drones', *Comput. Commun.*, vol. 148, no. September, pp. 208–214, 2019, doi: 10.1016/j.comcom.2019.09.018.
[32] A. Abdelmaboud, 'The Internet of drones: Requirements, taxonomy, recent advances, and challenges of research trends', *Sensors*, vol. 21, no. 17, 2021, doi: 10.3390/s21175718.
[33] M.-A. Lahmeri, M.A. Kishk and M. -S. Alouini, 'Artificial intelligence for UAV-enabled wireless networks: A Survey', *IEEE Open J. Commun. Soc.*, vol. 2, pp. 1015–1040, 2021, doi: 10.1109/OJCOMS.2021.3075201.
[34] Z. Jiao *et al.*, 'A Deep Learning Based Forest Fire Detection Approach Using UAV and YOLOv3', *Proc. 1st Int. Conf. Ind. Artif. Intell.(IAI)*, 2019, pp. 1–5, doi: 10.1109/ICIAI.2019.8850815.
[35] C. Chola, J.V. Benifa, D.S. Guru, A.Y. Muaad, J. Hanumanthappa, M.A. Al-antari, H. AlSalman, and A.H. Gumaei, 'Gender identification and classification of drosophila melanogaster flies using machine learning techniques', *Comput. Mathematical Methods Med.*, vol. 2022, pp. 1–9, 2022, 10.1155/2022/4593330.
[36] J.B.R. Rose, S.G. Natarajan, and V.T. Gopinathan, 'Biomimetic flow control techniques for aerospace applications: a comprehensive review', *Rev. Environ. Sci. Biotechnol.*, vol. 20, pp. 645–677, 2021. 10.1007/s11157-021-09583-z.

2 Modeling, Simulation, and Analysis Hybrid Unmanned Aerial Vehicle with C-Wing

S. Praveena, P. Karthikeyan, N. Srirangarajalu, and M.K. Karthik

CONTENTS

DOI: 10.1201/9781003252085-2

2.1 INTRODUCTION

A UAV is a type of aircraft and operated by without human. The first practice of the unmanned aerial vehicle concept happened in Austria as a balloon carrier in the year of 1849. They aimed the balloon carriers to the surrendering places; some of the balloons reached their aims but most of them changed their direction due to the wind. The major development of UAV was introduced in the 1900s. The UAVs were mainly used for training the army to practice the flying target.

After many decades of evolution, now UAVs are being used in common practices, starting from video coverage in marriage functions to surveillance in defense. Now the technology has been stepped into connecting and coordinating the UAVs with a base station using cloud storage known as Internet of Drones. It is leading to the development of precise positioning, real-time communication, and seamless coverage. The architecture of an IoD is shown in Figure 2.1. This chapter intends to aid the world of Internet of Drones by increasing the flight time of the UAV systems via modifying the winglet.

UAVs are majorly categorized into two types. They are rotary type (drone) and fixed-wing type (RC plane). Every type has its merits and demerits. While taking demerits, fixed-wing planes need a runway system for takeoff and landing. On the other hand, rotary types have limited endurance and range of operation. To overcome the drawbacks of conventional UAV, hybrid unmanned aerial vehicles were introduced. They can take off and land vertically, known as vertical takeoff and landing type (VTOL) and can move forward with high speed.

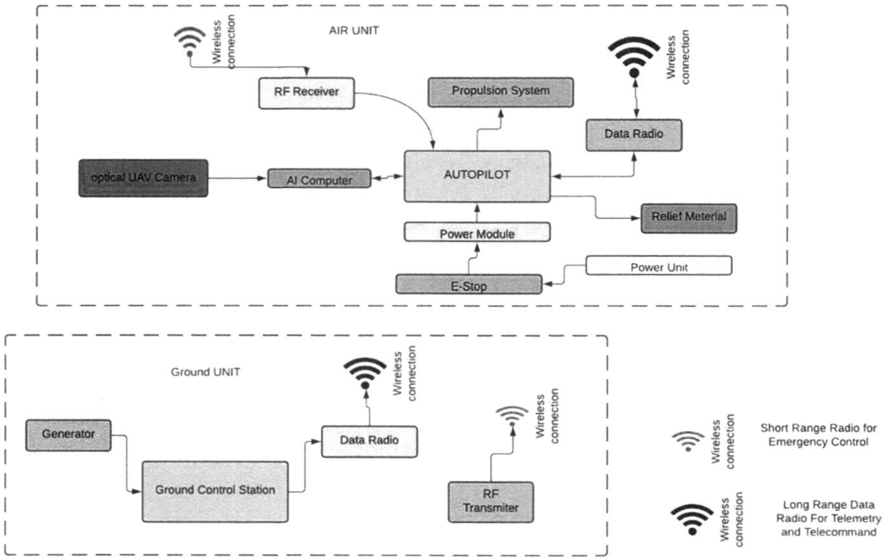

FIGURE 2.1 Architecture of IoD.

2.2 LITERATURE REVIEW

This section deals with the literature review on hybrid UAVs with a C-winglet. The literature has been classified as system development and control strategy. The system development part gives an introduction on kinematics equations and model structure while the control part deals with parameter estimation and introduction to different control strategies used in existing models.

KrooI et al. (2005) introduced various nonplanar wing concepts such as box wing, joined wing, C-wing, biplane, and wing with winglets. These concepts were used for reduction of vortex drag, which is also known as induced drag [1]. Induced drag is an opposing force for the aircraft's forward movement, which also depends on the lift of the aircraft: when lift increases, induced drag is also increased. The reduction of induced drag significantly contributes on fuel consumption that can be achieved by modification of wing tip. It will be described in this paper.

Takaaki Matsumoto et al. (2010) talked about a novel solution for disturbances occurring during hover. They applied various control strategies for reducing these errors. First, a conventional quaternion feedback controller was used for stability [2]. This controller works based on controlling the tilt twist angle based on the error in the Z axis. But it failed because of large attitude error caused by blast wind or bird strike.

Harijono Djojodihardjo et al. (2013) designed a joined-wing business jet that was lighter than the reference jet by 5.0716%. It achieved 1.3629 times of lift and reduction of 3.5% of total drag. The major advantage of reducing the wingtip vortices results in mitigation of wake turbulence [3]. Natassya Barlate Floro da Silva et al. (2013) introduced Tiriba, also known as TLUAV (tail-sitter), with modern characteristics but had a drawback of an undeveloped autopilot system. This type of UAV acted as a glider, hence the takeoff and landing require manual launching [4]. The main advantage of this model is gathering the geo-position referenced images and following the way points given from the base station.

Ugur Ozdemir et al. (2013) explained designing (weight estimation, wing geometry, and propulsion system) and testing of a TURAC VTOL UAV and compared it with other UAVs. That project involved designing two variants A and B with a different shaped fuselage and blended winglet with swept wing. It didn't have a tail structure for moment balancing. To overcome this problem, reflex-type airfoil was used for better aerodynamic performance. For maneuvering, two tilted motors were used for cruising performance and one electric fan motor was used for lifting the model. The main drawbacks of that project was the model had additional weight due to the VTOL system and it was difficult to control due to a high thrust to weight ratio compared to other conventional UAVs [5]. But an autopilot system might resolve those problems.

Ertugrul Cetinsoy et al. (2015) projected an improved design and control system for hybrid UAVs through the Newton-Euler method [5]. They used various controllers such as inner-loop controller, back stepping controller, nested saturation altitude controllers, singular perturbation theory-based controllers, PID controller, P controller, and LQI controller for various operations like hover mode, horizontal flight mode, transition mode, quad tiltrotor, quad tilt wing, yaw motion, and roll-pitch dynamics. This study gave different solutions when the vehicle suffered from various disturbances.

Sahana D S et al. (2015) conducted CFD analysis on various design parameters like h/b ratio and cant angle. It is the most important variable, which directly influences the overall efficiency. From their analysis, h/b = 1 and cant angle = 10° was arrived for optimum induced drag for various flight conditions [6]. Suresh C et al. (2016) used two approaches. One was CFD analysis and another one was wind tunnel test of the C-wing; 100° cant angle gave improved performance and gave optimum C_L/C_D ratio. By using a C-wing, induced drag was reduced, and wingtip tip vortices was converted as additional thrust [7]. Michael Becker et al. (2016) presented a hybrid unmanned aerial vehicle used for an ISR (intelligence, surveillance, and reconnaissance) mission. Here, COTS (commercial off the shelf) is for reducing cost effects and easy reconstruction [8]. Tiltrotor was used for the vertical and horizontal flight that vehicle was controlled by the Android device. By combining both quadcopter and fixed-wing UAV, more cruise speed, more payload capacity, and more endurance was achieved.

Yijie Ke et al. (2016) proposed a J-LION UAV design with a morphing wing. First, the model was designed in CAD software and analyzed using ANSYS ICEM. Unstructured mesh was done with superior quality of 0.2 [9]. This meshing creates 1.86 million elements, and the meshing model was imported to ANSYS fluent. For analyzing purposes, the Spalart-Allmaras viscous model was selected due do its fast-running capability and it obeyed the Navier-Strokes equations. From the results, the lift produced by the fuselage is 30% of the total drag and 10 N lift produced when using an extensible wing.

Ugur C et al. (2017) discussed the design parameters and selection criteria of airfoil for the remotely controlled fixed-wing aircraft [10]. The aircraft was designed by primarily considering the aerodynamic factors such as lift, drag, weight, and thrust. The weight of the vehicle was significantly reduced using a balsa-type wing to give higher lift, higher thrust, and lower drag.

Haowei Gu et al. (2017) proposed a design, verification, and validation of hybrid UAVs. They described how to design a model for communicating between two models and to a ground station. The vehicle designed and autonomous flight is achieved by controller design [11]. The position signal was detected from the base stations or from the waypoints, so that the attitude and total thrust required were calculated. Then it was mapped to the position controller to achieve operations like vertical takeoff, transition, and cruise. Finally, it was fed to the VTOL mixer for complete controlling.

Yijie Ke et al. (2018) proposed a model with vertical takeoff and vertical landing. The control system was designed with stable dynamic and flight control techniques, which was arrived from experimental results [12]. A novel tail-sitter hybrid UAV had good aerodynamic and multiple flight modes. Janith Kalpa Gunarathna et al. (2018) explained Sky Scout fixed-wing drone development and experimental test result. This paper showed the result of a quad plane's superior flying quality. The experimental data were given by testing at 53% for throttle hovering, cruise speed about 13 m/s, and 55% for cruise throttle [13]. Flight dynamics were examined using MATLAB Simulink software. The effects of wind were solved by a feedback control loop.

Enrique Paiva et al. (2018) proposed that hybrid topology will have the vertical takeoff and landing (VTOL) ability as well as better efficiency during the flight [14–18]. They applied a linear controller to declare the performance of the anticipated design experimentally. Based on the results, they concluded that both the fixed-wing UAV (RC plane) and rotary-wing UAV (drone) have their merits and demerits on their own design.

By combining both designs, the drawbacks such as lower cruise speed, requirement of runway, lower endurance, payload capacity, and lower range caused by rotary-wing and fixed-wing UAVs can be resolved. Alternatively, when commercial aircraft flies at a high speed, the vortices are formed at the tip of the winglet due to the interaction of chord-wise flow and span-wise flow. These vortices create downwash, which will cause the production of induced drag. During low-speed conditions such as takeoff and landing, the induced drag contributes more to the total drag. The non-planar wing configuration can maximize the lift and reduce the induced drag in wingtips by reducing vortex strength. This proposed design was quite complex to implement directly at the commercial aircraft on safety and consideration of cost. Hence, this proposed work is executed in UAVs only.

2.3 METHODOLOGY

The methodology of fabrication of hybrid UAV is shown in Figure 2.2. Initially, the UAV design was done in solid works and was exported to the numerical analysis platform to understand the aerodynamic behavior of the system. Then, the optimized design model was exported to the virtual simulation and control platform. Based on the numerical simulation and virtual simulation, the design and a control system were finalized before the fabrication. During the fabrication and integration phase, the selection of structural components, motors, controllers, battery, and support links were finalized based on components' cost and availability.

2.3.1 Design Selection

Based on the references [11], a dual system unmanned aerial vehicle is selected for developing a model. Because it has independent propulsion systems, that means

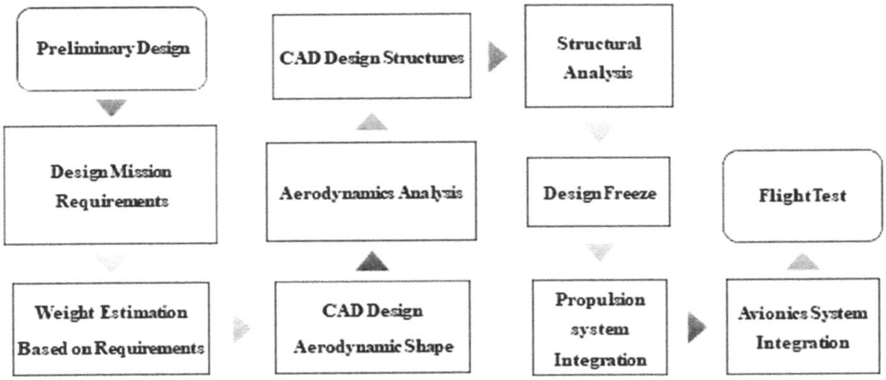

FIGURE 2.2 Methodology to develop a UAV.

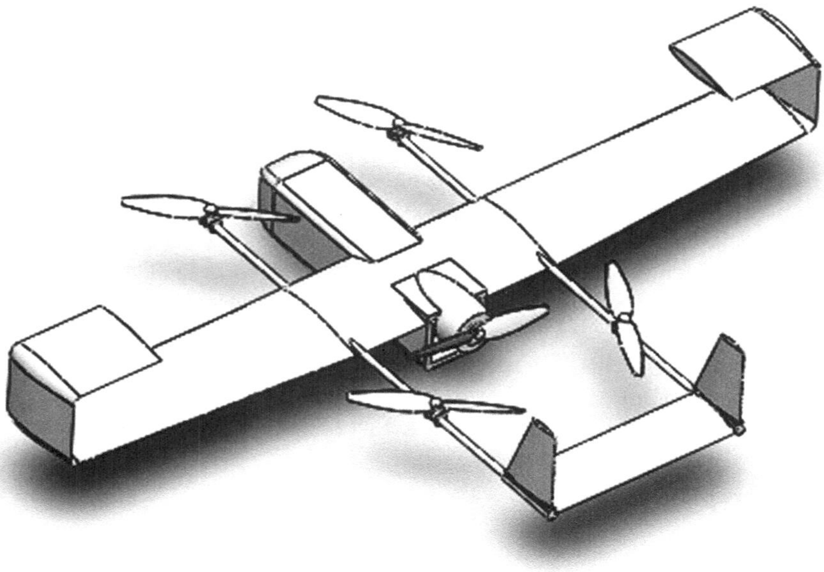

FIGURE 2.3 Proposed hybrid UAV.

four 920 kV rating motors for hovering and 2,300 kV rating motors for level flight and easy to manufacture, control, and maintain. Figure 2.3 shows the proposed hybrid UAV. The dual system unmanned aerial vehicle with a C-winglet is modeled by using CAD software. Each part file is first modeled and then they are assembled to form the complete device. It details the chord length, span length, motor mounting distance, vertical stabilizer length, horizontal stabilizer length, etc.

2.3.2 Selection of Airfoil

To construct the vehicle, NACA 64215 is the airfoil taken as a reference from the aircraft Airbus A300-600R due to the significance of supercritical airfoil. NACA 64215 coordinates are used for drawing airfoil shape, and it is a cross-section of the wing. It is used for making aircraft wings, horizontal stabilizers, and vertical stabilizer. This airfoil has a maximum chord thickness of 15 mm.

2.3.3 Selection of Motors

The approximate weight of the hybrid UAV, including all other components, is 1,632 grams. So required thrust for one lifting motor with considering a factor of safety 1.5 is 612 g. For that reason, a DJI 2212 920 kV BLDC motor selected.

$$Thrust = \frac{Total\ \ weight\ \ \times\ \ FOS}{number\ \ of\ \ motor} = \frac{1632 \times 1.5}{4} = 612g$$

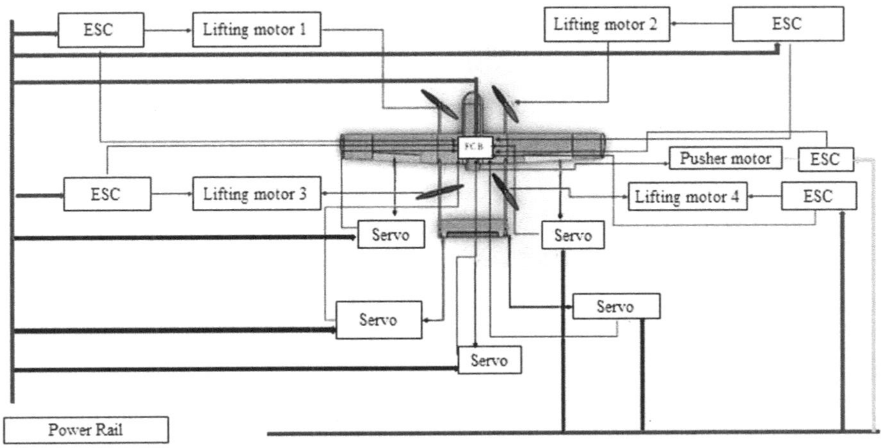

FIGURE 2.4 Schematic diagram of UAV components.

2.3.4 Schematic Diagram of UAV Components

The line diagram of our project model is shown in Figure 2.4; here, the red colour line indicates the connection of power rails from the battery to brushless direct current (BLDC) motors, electronic speed controller (ESC), and flight controller. The green line indicates the servo connections for secondary control surfaces like the elevator, aileron, and rudder to the flight control board. The violet line indicates the signal wire from the flight controller to drivers. This dual system UAV is made up of Styrofoam and aluminum for economic conveniences and weight considerations. The model consists of four lifting motors, one pusher motor, five ESCs, power distribution board, battery, transmitter, receiver, and flight control board. The battery and ESCs are connected to the power distribution board using soldering. Then, power from ESCs will be distributed to BLDC motors and servos. Here, servo motors are used for actuating control surfaces like aileron, elevator, and rudder. The signal wires from ESCs are connected to the flight control board, and the flight control board powers the receiver. ESCs function as drivers for BLDC and servo motors because it is built with a battery elimination circuit. A BEC (battery elimination circuit) acts like a regulator.

2.3.5 Specifications of UAV

1. Wing Span = 1,000 mm
2. Selected supercritical airfoil is 64215
3. Vertical Tail Distance = 65 mm
4. Horizontal Stabilizer Distance = 155 mm
5. Height of Vertical Winglet = 59.66 mm (23% of span from "AIAA journal" on 2014)

6. Height of Horizontal = 78.97 mm (26% of span from "AIAA journal" on 2014)
7. Lifting Motor Mounting Distance = 600 mm (vertical).
8. Kv Rating of BLDC Motor = 920 kV
9. Battery Capacity = 2,200 mAh

2.3.6 Component Selection and Weight Estimation

To fabricate and integrate the model, components to be selected and the self-weight estimated in Table 2.1 is used to calculate the self-weight and payload.

Table 2.2 indicates the specifications of software and hardware components.

TABLE 2.1
Elements in UAV and Their Weight

Device	Quantity	Weight (g)
Power distribution board	1	8
Battery	1	175
Flight control board	1	30
Lifting motors	4	224
ESC for lifting motors	4	92
Pusher motor	1	30
ESC for pusher motor	1	23
Frame	N-A	700
Wing	N-A	350
Total weight	N-A	1,632

TABLE 2.2
Specifications of Software and Hardware Components

Description	Specifications	Quantity
Lifting Motors	DJI 2212 920 kV BLDC motor	4
Pusher Motor	RS 2205 2,300 kV BLDC	1
Electronic Speed Controller	30A ESC	6
Flight Control Board	2.4.8 Pixhawk Flight Controller	1
Battery	Orange 2,200 mAh 3S 30C/60C Lithium polymer battery Pack (LiPo)	1
Transmitter	FS-TH9X 2.4 GHz 9CH	1
Receiver	FS-IA10B	1
Power Distribution Board	100 A Multirotor ESC Power Distribution Battery Board	1

2.3.7 Selection of Winglet

In recent times, fuel economy has been the main important parameter in the air transportation industry. One of the ways to reduce fuel consumption is the reduction of drag. For example, the commercial aircraft of airbus A-340 operates under the decrease of 1% of the total drag saves 400,000 liters of fuel and 5000 kg of insalubrious emissions per year. Lift and drag are considered significant components of total aerodynamic forces. Aerodynamic drag typically consists of friction drag and induced drag. Friction drag is resolved by the state of the boundary layer (laminar, transition, or turbulent) and does not vary grandly at subsonic. It contributes 40%–45% of the total drag of an airplane. For takeoff or landing, induced drag comprises up to 90% having substantial impingement on aircraft performance. Conventional winglets make utmost drag hold back available and get a better lift to drag ratio under cruise conditions. During takeoff and landing, these winglets don't give optimum results. For these reasons, non-planar winglets are introduced for reducing the contribution of induced drag to the total drag. The reduction of drag in aircraft will directly affect operational cost optimization and an indirect effect on noise and emission reduction. At the time of lift production, the 1% reduction of drag will enhance lift to drag ratio at cruise phase, resulting in improved aerodynamic efficiency.

2.4 RESULTS AND DISCUSSION

2.4.1 ANSYS Simulations and Results

ANSYS simulation is used to determine various aerodynamics factors such as lift coefficient, drag coefficient, moment, velocity, and pressure contour over the aircraft model.

2.4.1.1 Fluent Analysis for Hybrid UAV with C-Winglet

An external geometry file (.IGS) was imported and generated. The model is covered with cubic enclosure. Each face and model was named as inlet, outlet, external wall, and model. Finally, a model is subtracted from the enclosure for optimum flow over a model. Figure 2.5 shows the image of meshing of a hybrid UAV with a C-winglet.

Required conditions would be set, and meshing has been done after some more times depending on the mesh type. Accuracy and better results are obtained using

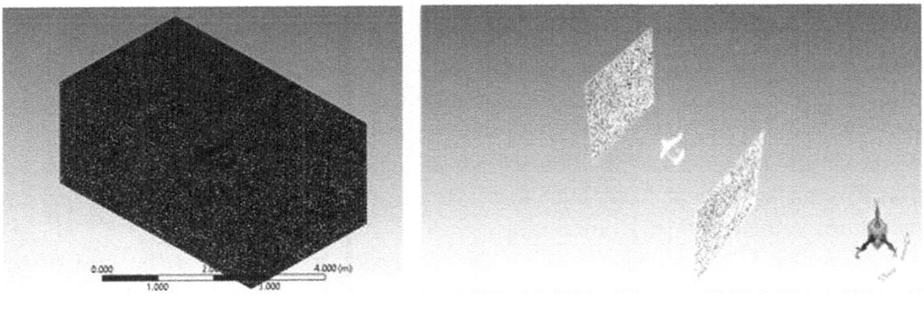

FIGURE 2.5 Meshing of model.

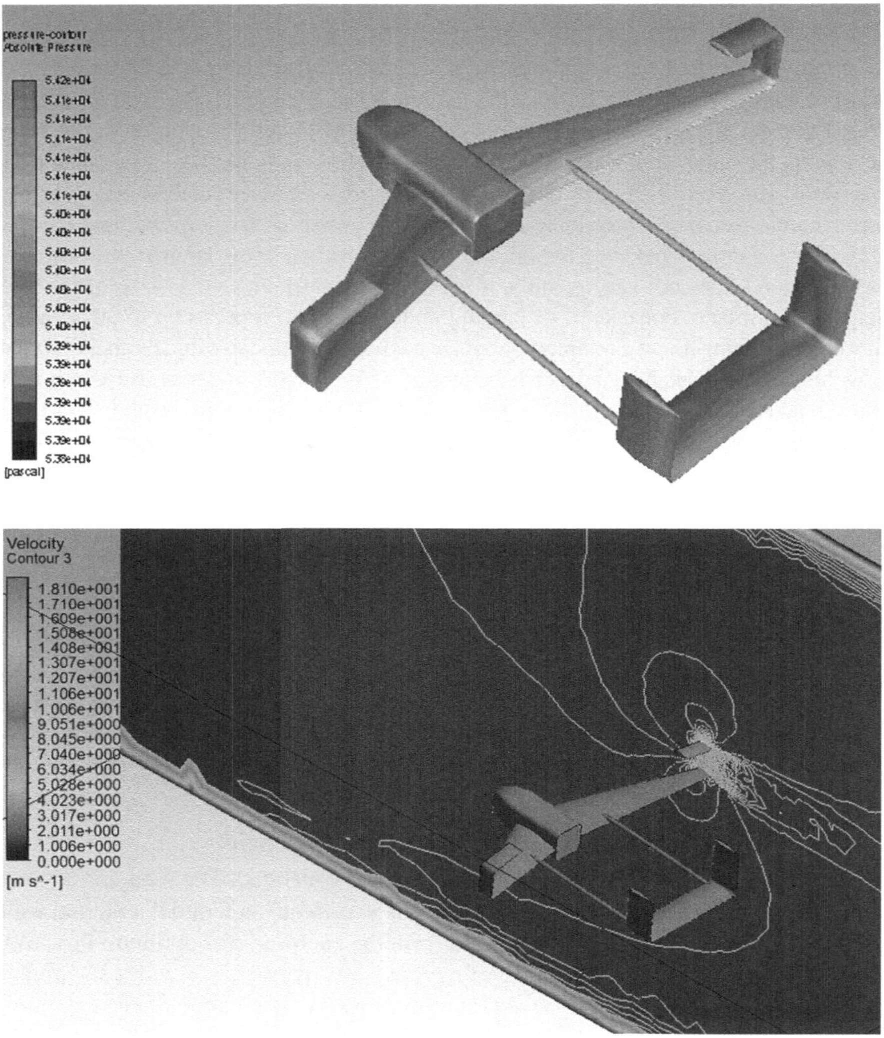

FIGURE 2.6 Pressure contour over model and velocity contour at tip of the C-wing.

mesh. A K-Omega turbulent model solution setup is selected. Figure 2.6 shows the information of absolute pressure over the model and velocity in a three-dimensional representation of the wing with the C-winglet model. Here the red colour indicates the maximum pressure and the blue colour indicates the low pressure in the contour plot. The airflow starts from inlet to outlet. From fluid flow analysis, we get the maximum velocity of 18.10 m/s.

The vortex is formed at the wing tip and wake are formed at the mid of the horizontal winglet and wing. Due to these actions, further lift will be created at the wing.

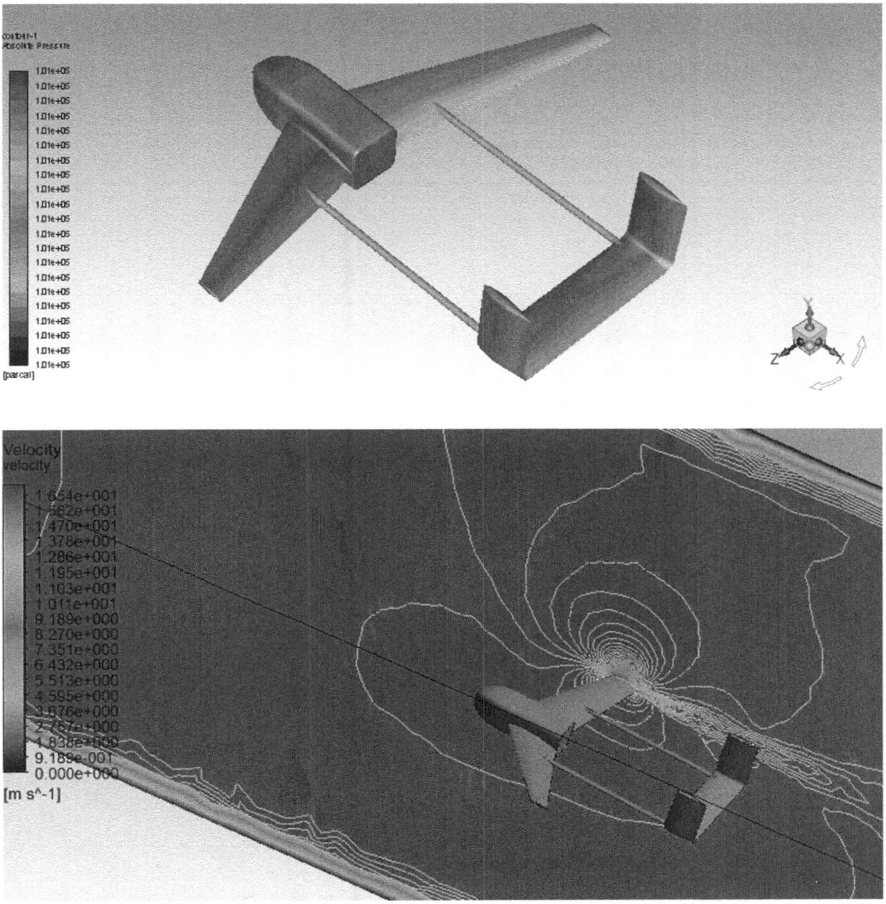

FIGURE 2.7 Pressure contour over the model and velocity contour on the plane.

2.4.1.2 Fluent Analysis for Hybrid UAV without C-Winglet

Absolute pressure information around the hybrid UAV without the winglet and the maximum velocity experienced by the model in the plane is around 16.54 m/s, as shown in Figure 2.7. Here, the plane is drawn from the Z axis at the distance of –0.3 m.

2.4.1.3 Fluent Analysis for Wing without Winglet for Various AOA

Table 2.3 shows that the value of the coefficient of lift and coefficient of drag at 5° angle of attack for a wing without winglet.

Figures 2.8 shows the pressure contour and velocity contour image for a wing without a winglet.

TABLE 2.3
C_L and C_D Value for Wing Without Winglet

Wing without winglet	Coefficient of lift	Coefficient of drag
	1.125 e^{-2}	1.264 e^{-3}

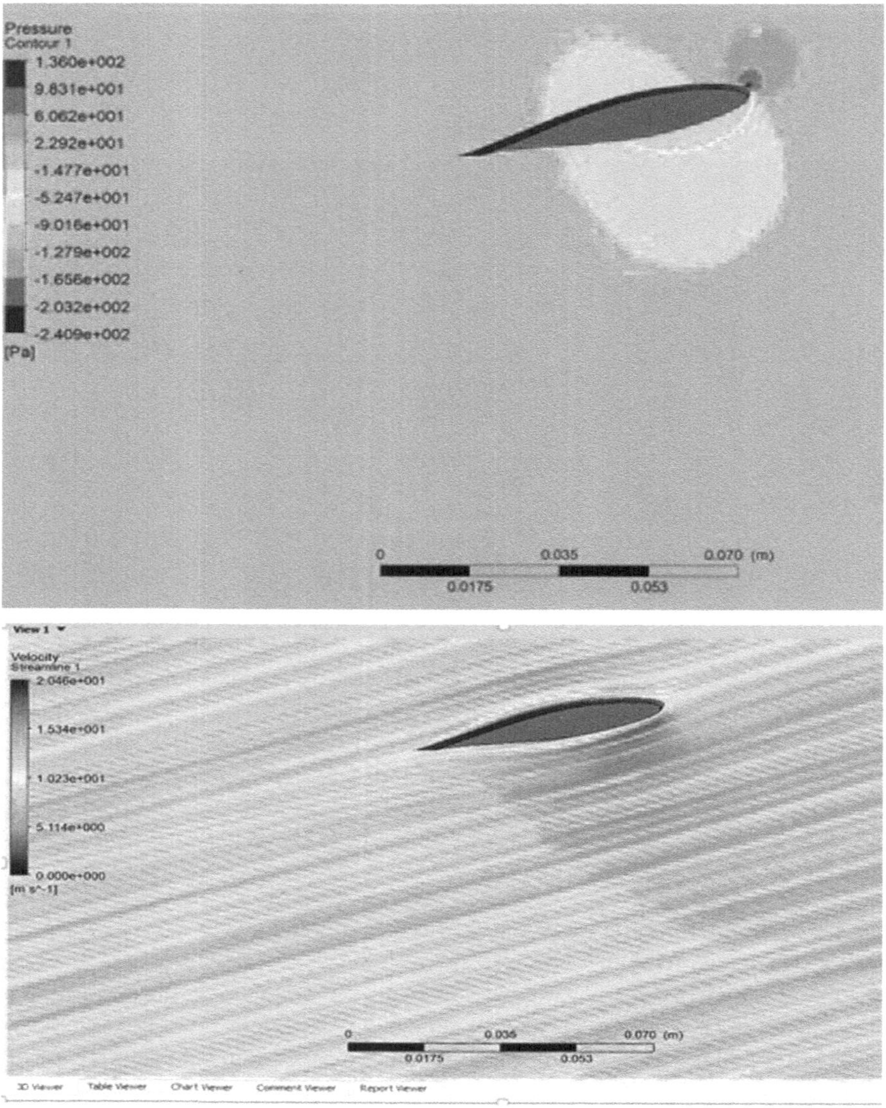

FIGURE 2.8 Pressure contour and velocity streamline for a wing without a winglet.

2.4.1.4 Wing Area Calculations for a Wing without a Winglet

$$\text{Area for rectangular wing} = s \times c$$

Where,
- s – Wing span
- c – Chord length

$$\text{Wing area for rectangular wing} = 280 \times 170$$
$$A_1 = 47600 \text{ mm}^2$$

$$\text{Wing area for tapered wing} = [0.5[Cr + Ct] \times S]$$

Where,
- C_r – Root chord length
- C_t – Tip chord length
- S – Span length

$$\text{Wing area} = (0.5[170 + 140] \times 360)$$
$$A_2 = 55800 \text{ mm}^2$$

$$\text{Total Area of the commercial wing} = [A1 + 2(A2)]$$
$$= 159200 \text{ mm}^2$$

$$\text{Aspect ratio of the commercial wing} = \frac{b^2}{s}$$
$$= \frac{(wing\ span)^2}{wing\ area}$$
$$= \frac{(1000)^2}{159200}$$

$$\text{A.R} = 6.28$$

$$\text{Wing loading } \frac{w}{s} = \frac{Body\ mass}{wing\ area\ in\ mm}$$
$$= \frac{1.5}{159200}$$
$$\frac{w}{s} = 9.42 \text{ kg/m}^2$$

$$\text{Induced drag co-efficient for without winglet } C_{Di} = \frac{C_L^2}{\pi \cdot e \cdot AR}$$

$$= \frac{0.01258^2}{(\pi \times 1.04 \times 6.281)}$$

$$C_{Di} = 6.176 \times 10^{-6}$$

$$\text{Induced drag for wing without winglet } D_i = [0.5 \times \rho \times v^2 \times s \times C_{Di}]$$

$$= 0.5 \times 1.225 \times 15^2 \times 0.1592 \times 6.176 \times 10^{-6}$$

$$D_i = 1.35499 \times 10^{-4}\,\text{N}$$

2.4.1.5 Induced Drag for a Wing with a C-Winglet for Various AOA

Table 2.4 shows the results of aerodynamic parameters for a wing with a C-winglet.

Figure 2.9 shows that the image of pressure and velocity contour for a wing with a C-winglet.

2.4.1.6 Induced Drag Calculations for a Wing with a C-Winglet

$$\text{Area for vertical winglet} = \text{s} \times \text{c}$$

Where,
- s – Wing span
- c – Chord length

$$\text{Area for vertical winglet} = 115 \times 140$$

$$A_3 = 16100 \text{ mm}^2$$

$$\text{Area for horizontal winglet} = \text{s} \times \text{c}$$

Where,
- s – Wing span
- c – Chord length

$$\text{Area for horizontal winglet} = 130 \times 140$$

$$A_3 = 18200 \text{ mm}^2$$

TABLE 2.4
C_L and C_D Value for a Wing with a C-Winglet

Wing with C-winglet	Coefficient of lift	Coefficient of drag
	1.234×10^{-2}	1.645×10^{-3}

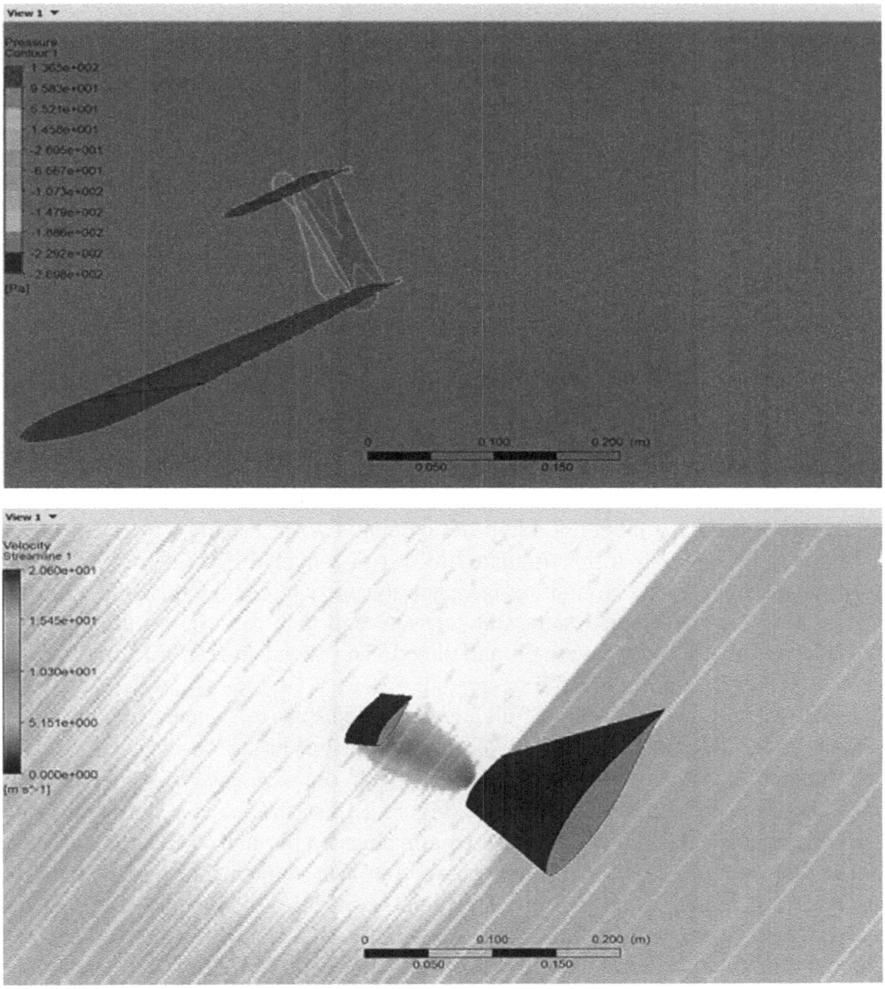

FIGURE 2.9 Pressure contour and velocity streamline for a wing with a C-winglet.

$$\begin{aligned}\text{Total area for C-winglet} &= \text{Total Area of the commercial wing} \\ &\quad + \text{Area for horizontal winglet} \\ &\quad + \text{Area for horizontal winglet} \\ &= 193500 \text{ mm}^2\end{aligned}$$

$$\begin{aligned}\text{Aspect ratio of the C-wing, A.R} &= (\text{b2/s}) \\ &= (\text{wing span})^2/\text{wing area} \\ &= (1245)^2/193500 \\ \text{A.R} &= 8.01047\end{aligned}$$

$$\text{Induced drag co-efficient for with C-winglet } C_{Di} = \frac{C_L^2}{\pi \cdot e \cdot AR}$$
$$= \frac{0.012352^2}{(\pi \times 1.30 \times 8.01047)}$$
$$C_{Di} = 4.663 \times 10^{-6}$$

$$\text{Induced drag for C-wing} = [0.5 \times \rho \times v2 \times s \times C_{Di}].$$
$$= [1.5 \times 1.225 \times 15^2 \times 0.1935 \times 4.663 \times 10^{-6}].$$
$$= 1.2434 \times 10^{-4}\text{N}$$

2.4.2 Physical Modeling in a Simulation Platform

The dual system unmanned aerial vehicle would be designed by CAD software and simulated via virtual platforms.

Simulink is used for simulating and analyzing the various aerospace vehicle dynamics. It will give practical feedback about the vehicle before we go for experimental design. MATLAB SIMULINK is the virtual library. Figure 2.10 shows that the imported simscape model of a hybrid unmanned aerial vehicle with a C-winglet.

Figure 2.11 shows that the simulated model of a hybrid VTOL UAV system.

If the response of the simulation is very poor, we can change the model design. First, a model was designed using 3D software and then Simulink multibody created and a .xlm file was generated and saved. The saved model will be imported with the data type smimport<file name>.

It will explain the interaction of high-level components and understanding the behaviour of the design. Investigation of specific design during design phase reveals rather than the construction phase overall cost effect. The advantage of a simulator is to know about the problem due to obstacles. Figure 2.12 shows the closed system model of an unmanned aerial vehicle. It would be implemented through MATLAB Simulink.

The imported model is considered as a stationary object, because there is no movement of model in its inertial directional axis. In experimental setup, an unmanned aerial vehicle is actuated by a brushless DC motor and controlled by a pixhawk flight controller and electronic speed controller. Similarly, in the software,

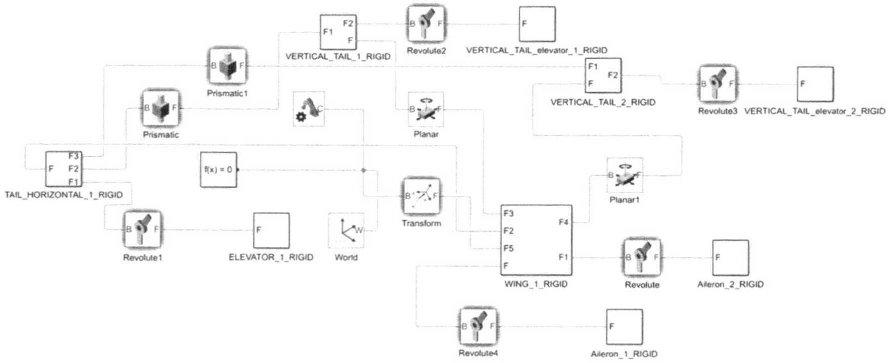

FIGURE 2.10 Physical model of hybrid UAV.

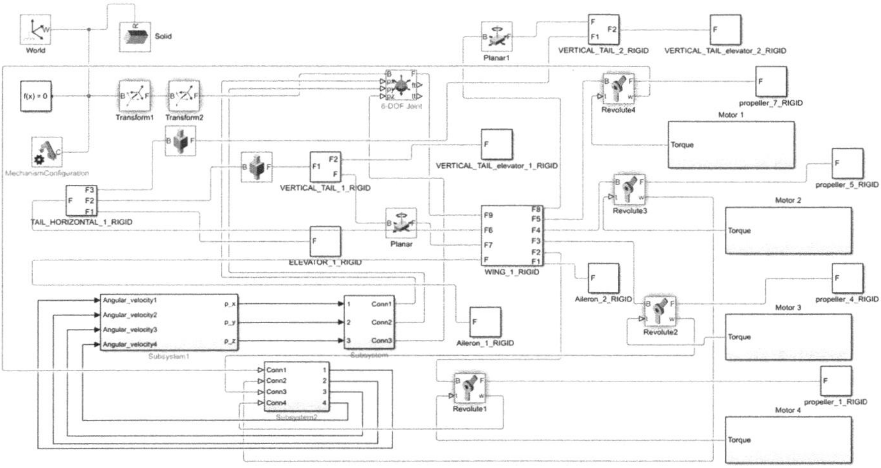

FIGURE 2.11 Simulated model.

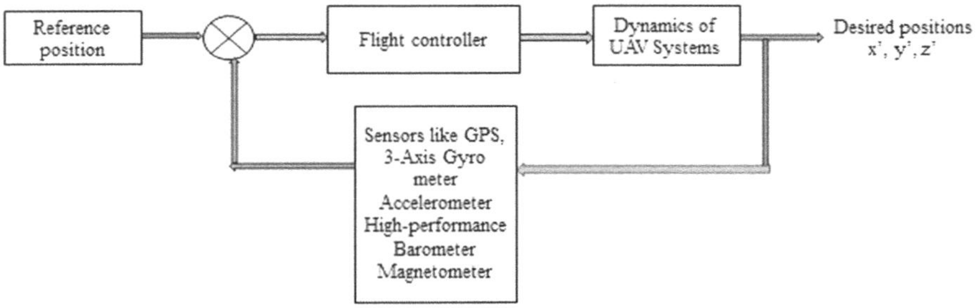

FIGURE 2.12 Block diagram for closed loop control system of UAV.

a propeller is actuated by a DC motor and controlled by dynamic equations. A virtual illustration of a real-world system consists of software and hardware. Mathematical modeling is used for drive software components.

2.4.3 UAV Dynamics (U_1 [N])

The hybrid UAV consists of both quadcopter dynamics and fixed-wing dynamics. The drone consists of four lifting motors for hover, roll, pitch, and yaw. Four fixed pitch propellers are attached to the brushless DC motors. Two rotors rotate in the clockwise direction and two rotors rotate in the counterclockwise direction. When all these motors rotate at the same speed and torque, the quadcopter achieves a hover (throttle speed). Equation (2.1), equation (2.2), equation (2.3), and equation (2.4) are used to derive the dynamic state of the unmanned aerial vehicle are as follows:

$$\ddot{X} = (sin\, sin\, \alpha\, sin\, sin\, \beta + cos\, cos\, \alpha\, sin\, sin\, \Upsilon\, cos\, cos\, \beta) \times \frac{U_1}{m} \quad (2.1)$$

$$\ddot{Y} = (\alpha\, sin\, sin\, \beta + sin\, sin\, \alpha\, sin\, sin\, \theta\, cos\, cos\, \beta) \times \frac{U_1}{m} \quad (2.2)$$

$$\ddot{Z} = -g + (cos\ \ cos\, \Upsilon\, cos\, cos\, \beta) \times \frac{U_1}{m} \quad (2.3)$$

$$\Omega = (-\Omega_1 + \Omega_2 - \Omega_3 + \Omega_4) \quad (2.4)$$

Where,

X, Y, Z – positions
α – Yaw angle
β – Roll angle
ϒ – Pitch angle
U_1 – Throttle speed
m – Mass
b – Thrust factor.

Figure 2.13 shows that the result of a simulation done by MATLAB Simulink. By doing this simulation, the dynamics of the UAV can be predicted.

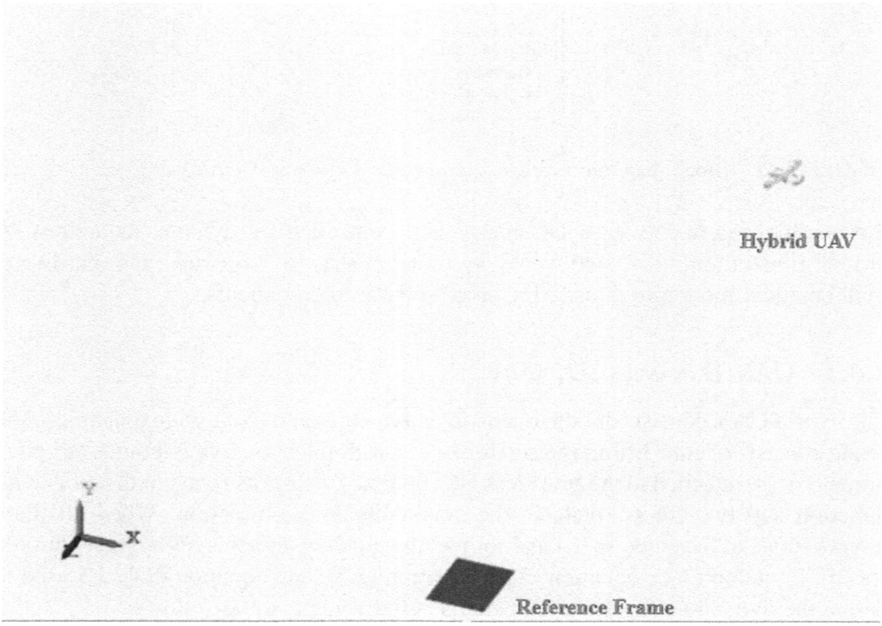

FIGURE 2.13 Simulation result.

For the fixed-wing unmanned aerial vehicle dynamics, we want to find lift coefficient and drag coefficient. C_L and C_D values are determined by using ANSYS software.

2.5 EXPERIMENTAL SETUP

A dual system UAV is the combination of drone and fixed-wing unmanned aerial vehicle.

2.5.1 Rotary-Wing UAV

An aluminium frame is selected for making a rotary-wing UAV base; 3.5 mm diameter holes for mounting a brushless DC motor are created using a drilling process with required balancing distance. A motor holder and 3 mm screw are used for BLDC motor mounting. Aluminium square tubes are attached through L clamps. Two stages are formed by a 1 mm aluminium plate. Aluminium plates are bent with a shearing machine. Figure 2.14 shows the design of the frame setup with BLDC motors and the image of wing and tail configuration that would be made up of extruded Styrofoam. It is made up of aluminium for cost efficiency and weight to strength ratio considerations. Four electronic speed controllers are soldered with a power distribution board, and they are fixed with an aluminium frame. Next, a battery connector is soldered with a power distribution board, and it is attached through a double-sided cello tape on the frame.

Here, two motors have clockwise rotation and another two must counter-clockwise rotation. Fourteen AWG wires are used for connecting electronic speed controllers with brushless DC motors through bullet connectors. Required propellers are attached to the BLDC motor shaft. It has a 10 mm length and 4.5 mm pitch. Pixhawk flight controller is connected to the ESC and fixed in the frame over double-sided cello tape and the receiver is connected to the flight controller and gets a signal from the controller via the transmitter.

2.5.2 Fixed-Wing UAV

Styrofoam can be classified into two types. One is extruded and another one has expanded Styrofoam. Here, extruded-type Styrofoam is used for making a fixed-wing unmanned aerial vehicle because it has high strength to weight ration compared to other type of materials. Unsymmetrical airfoil NACA 64215 is used for developing a wing. First, airfoil coordinates are downloaded from the website and coordinates are multiplied with required dimensions using Excel. Finally, co-ordinates are imported to the solid works software to get the three-dimensional curve. Then it is converted into a two-dimensional curve by a solid works drawing. This two-dimensional airfoil is printed into the A4 sheet.

This airfoil is pasted at the breadth of the foam material. Nichrome wires have higher thermal conductivity compared to other materials. According to this reason, it is used for foam cutting. A 64 V power supply is connected to the

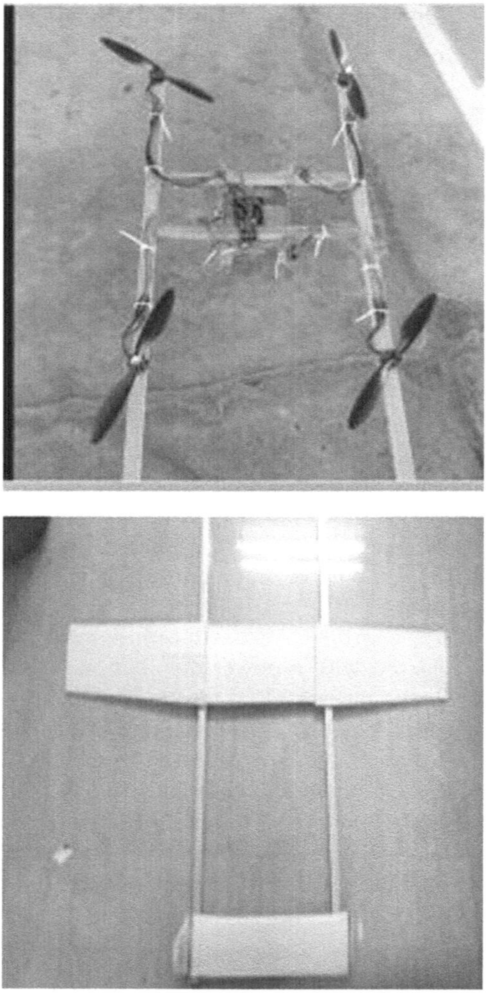

FIGURE 2.14 UAV frame and wing, tail configurations.

rheostat and Nichrome wire. When variable resistance is given via rheostat, the current over Nichrome wire causes thermal heating. This temperature is approximately 200°C, voltage is 20.7 V, and the current rate is around 0.9 A depending on the length of the wire. The middle of the aircraft is occupied with a rectangular wing that has the dimension of 170 mm. The other two sides of the aircraft have tapered wings with the dimension of root chord length of 170 mm and tip chord length of 140 mm. Similarly, the vertical stabilizer and horizontal stabilizer are made by way of manual foam cutting. Figure 2.15 shows the aircraft wing, vertical stabilizer, and horizontal stabilizer made by Styrofoam and UAV at flying mode.

FIGURE 2.15 UAV at flying mode.

2.5.3 Software Setup

Quadcopter and RC plane dynamics are started and controlled by the ground control station of the mission planner. The first flight controller is connected to the mission planner software via a micro-USB cable. Then the required COM port is chosen for my autopilot on the home page of the software. It is shown in Figure 2.16.

COM5 is selected based upon our flight controller and the baud rate is set. The COM5 serial port is capable of transferring data of 115,200 bits per second. Go to initial setup and then obligatory firmware would be loaded into pixhawk 2.4.8 flight controller depending upon the frame's design.

Next, the chibiOS operating system was uploaded for smaller and efficient firmware. It is an embedded application for a 32-bit microcontroller. It has a kernel size with a maximum range of 5.5 KiB and a minimum range of 1.2 KiB and is also entirely subsystem activated using the STM32 cortex-M3 processor. Within a few seconds, the firmware was loaded and then the OK button was clicked. After loading the firmware, the connect button was hit to connect the autopilot to the

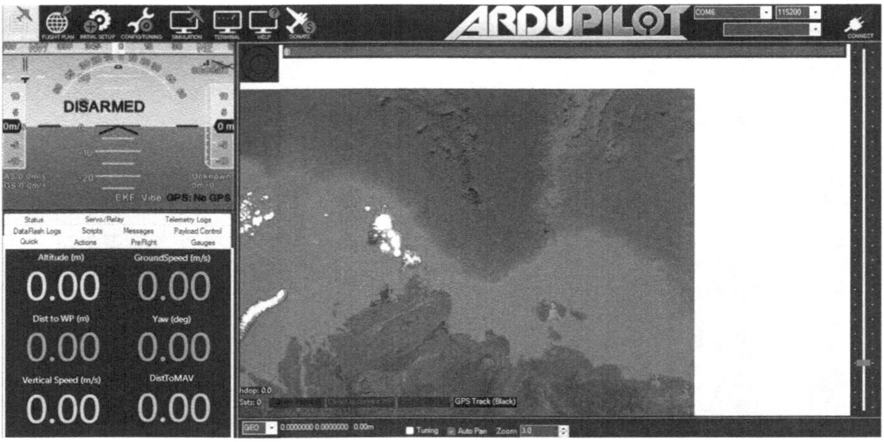

FIGURE 2.16 Mission planner software for flight controller.

software via the mav link communication protocol. Then, it was ensured the ardupilot firmware was installed and booted correctly. Pixhawk 2.4.8 has useful LEDs and a buzzer sound for indicating the state of the autopilot.

Autopilot has the processor of F7 or H7. It also has a CAN interface custom firmware that exists in two USB interfaces. One of the USB interface charity for MAVLINK connection and another one for SLCAN serial connection to the CAN interface for configuration and firmware updates. Pixhawk board has the serial port connection of SERIAL0 and SERIALX. Here MAVLINK USB interface act as a SERIAL0 port and SLCAN interface act as a SERIALX port. After installation of firmware, next we go to configuration or tuning. First, we do the compass calibration. Do the calibration near non-metallic objects because it causes incorrect calibration. Onboard calibration is more accurate than an off-board calibration because it is used to determine orientation, scaling, and offset. Click the onboard mag calibration's start button and the buzzer attached to the flight controller will give the single beep sound to indicate the ready to calibrate orientation such as front, back, left, right, top, and bottom. When a vehicle is rotated, the green bar in the mission planner software increases further and further unto the completion of calibration. After that, three beep sounds will be released; reboot the autopilot before arming the vehicle to get a better solution.

If calibration fails, an unhappy tone is emitted and the green bar is restarted to the left side and calibration restarts automatically if compass calibration still fails to increase the values of COMPASS_OFFS_MAX from 850 to 2,000 or 3,000. Onboard calibration is also done by the RC controller by throttling full up and right side yaw full for 2 seconds. If calibration fails, raise the throttle stick full up and left side yaw stick full for 2 seconds to cancel the calibrations. Next, we do the accelerometer calibration. While doing the acceleration calibration, ensure the correct level position for proper calibration. After completion of acceleration calibration, the mission planner will display "calibration successful." Then we do the radio calibration. This process is also common for RC planes. The RC transmitter does transition.

2.6 CONCLUSION

The hybrid UAV with a C-winglet was modeled using SOLIDWORKS with required dimensions. The control system and stability parameters for operation were found by analyzing the model dynamically in MATLAB SIMULINK. And then computational analysis was done using ANSYS fluent for finding aerodynamics parameters such as coefficient of lift and coefficient of drag for the model. The inference from the computational analysis for different wing structures was follows: UAV with a C-winglet has a lower induced drag than UAV without a C-winglet, which impacts the reduction of fuel usage and noise level and also improves the efficiency of the propulsion system. In the future, the drone itself will modify the parameters based on the data set in real time via IoD communication.

REFERENCES

[1] I. Kroo (2005). "Non-Planar Wings Concepts for Increased Aircraft Efficiency". *VKI lecture series on Innovative Configurations and Advanced Concepts for Future Civil Aircraft.*

[2] Takaaki Matsumoto, Koichi Kita, Ren Suzuki, Atsushi Oosedo, Kenta Go, Yuta Hoshino, Atsushi Konno, Masaru Uchiyama, et al. (2010). "A Hovering Control Strategy for a Tail-Sitter VTOL UAV that Increases Stability against Large Disturbance" IEEE International Conference on Robotics and Automation 1050–4729.

[3] Harijono Djojodihardjo and Kim Ern Foong (2013). "Conceptual design and aerodynamic study of joined–Wing business jet aircraft." *Journal of Mechanics Engineering and Automation*, 263–278.

[4] Natassya Barlate Floro da Silva, Kalinka Regina Lucas Jaquie Castelo Branco, et al. (2013). "A New Concept of VTOL as Fixed-Wing" International Conference on Unmanned Aircraft Systems (ICUAS) 978-1-4799-0815-8.

[5] Ugur Ozdemir, Yucel Orkut Aktas, Aslihan Vuruskan, Yasin Dereli, Ahmed Farabi Tarhan, Karaca Demirbag, Ahmet Erdem, Ganime Duygu Kalaycioglu, Ibrahim Ozkol, Gokhan Inalhan, et al. (2013). "Design of a commercial hybrid VTOL UAV system." *Journal of intelligent and robotic systems* 74, 371–393.

[6] E. Cetinsoy (2015). "Design and Control of a Gas-Electric Hybrid Quad Tilt-rotor UAV with Morphing Wing." International Conference on Unmanned Aircraft Systems (ICUAS) 978-1-5386-5587-0.

[7] D.S. Sahana and Abdul Aabid (2015). "CFD analysis of box wing configuration." *International Journal of Science and Research.* ISSN: 2319–7064.

[8] Michael Becker, David Sheffler, et al. (2016). "Designing a High Speed, Stealthy, and PayloadFocused VTOL UAV." IEEE Systems and Information Engineering Design Conference (SIEDS '16) 978-1-5090-0970-1.

[9] Suresh C., Ramesh K., Gobpinaath M., and Sivajiraja K. (2016). "Design and analysis of non planar wing in commercial aircraft." *International Journal of Innovations in Engineering and Technology*, 7(3), ISSN: 2319-1058.

[10] Yijie Ke, Hui Yu, Changkai Chi, Mingwei Yue, Ben M. Chen, et al. (2016). "A Systematic Design Approach for an Unconventional UAV J-Lion with Extensible Morphing Wings." 12th IEEE International Conference on Control & Automation (ICCA) 978-1-5090-1739-3.

[11] Haowei Gu, Ximin Lyu, Zexiang Li, Shaojie Shen, and Fu Zhang, et al. (2017). "Development and Experimental Verification of a Hybrid Vertical Take-Off and Landing (VTOL) Unmanned Aerial Vehicle (UAV)." 2017 International Conference on Unmanned Aircraft Systems (ICUAS) 978-1-5090-4496-2.
[12] Ugur C. Yayli, Cihan Kimet, Anday Duru, Ozgur Cetir, Ugur Torun, Ahmet C. Aydogan, Sanjeevikumar Padmanaban and Ahmet H. Ertas, et al. (2017). "Design optimization of a fixed-wing aircraft." *International Journal of Advances in Aircraft and Spacecraft Science,* 4: 65–80.
[13] Janith Kalpa Gunarathna, Rohan Munasinghe, et al. (2018). "Development of a Quad-rotor Fixed-wing Hybrid Unmanned Aerial Vehicle." Moratuwa Engineering Research Conference (MERCon) 978-1-5386-4418-8.
[14] S.N. Skinner and H. Zare-Behtash (2018). "Study of a C-wing configuration for passive drag and load alleviation". ISSN – 08899746. DOI: 10.1016/j.jfluidstructs.2017.12.018, Journal of Fluids and structures.
[15] Yijie Ke, Kangli Wang, and Ben M. Chen, et al. (2018). "Design and Implementation of a Hybrid UAV with Model-Based Flight Capabilities." *IEEE/ASME Transactions on Mechatronics,* 23(3): 1114–1125.
[16] Chuanguang Yang, Tielin Ma, Zilun Zhang, Gaoming Zhou, et al. (2019). "Research on Structural and Aeroelastic Design of a Hybrid Multi-rotor UAV" IEEE International Conference on Unmanned Systems (ICUS).
[17] Changsheng. (2020). "Hybrid Offline-Online Design for UAV-Enabled Data Harvesting in Probabilistic LoS Channel." DOI 10.1109/TWC.2020.2978073, IEEE Transactions on Wireless Communications.
[18] Yanchu Li, Qingqing Ding, Shufang Li, Stanimir Valtchev et al. (2021) "Optimal controller design for non-affine nonlinear power systems with static var compensators for hybrid UAVs". DOI 10.26599/TST.2020.9010058, ISSN 1007-0214.

3 Influence of Machine Learning Technique in Unmanned Aerial Vehicle

G. Prasad, K. Saravanan, R. Naveen, V. Nagaraj, R. Dinesh, and K. Anirudh

CONTENTS

3.1 INTRODUCTION TO UNMANNED AERIAL VEHICLE

A fundamental component of Industry 5.0 is Unmanned Aerial Vehicles (UAV) and Machine Learning (ML). A UAV (unmanned aerial vehicle) has the characteristics of lightweight, low cost, and the capacity to perform low altitude flights without a pilot. These abilities are in addition to the UAV being a duplicate of an airplane, portable, and capable of snapping items on the ground. Operators control the drone from afar, making it operational. Human operators don't need to be inside the UAV if it's controlled by a human pilot. Numbers of UAVs are likely to grow soon, according to present projections. UAVs come in numerous varieties, such as those that utilize traditional flight, rotary blade and hybrid Vertical Take Off and Landing (VTOL) technology. But the main elements of a UAV system are networks, aerial relays, and handheld terminals. To teach machines how to understand what they are

DOI: 10.1201/9781003252085-3

learning; data must be separated into two sets: the training and testing groups. Many algorithms are used for machine learning, including k-neural network, random forest, and support vector machines. The overall goal of this project was to compile the benefits, relevance, and potential future of various machine learning applications in the UAV. A detailed investigation was performed by searching years 2014–2021 academic journals in different digital databases.

3.2 UAV IN INDUSTRY 5.0

During the previous five years, the UAV sector has increased quite a bit. Based on Figure 3.1, total UAV revenue was estimated for 2016–2025 and appears to exhibit a rapid increase in the next several years. Using UAVs and machine learning in data mining can be effective in precision agricultural, infrastructure, and building construction projects. An example of the models that are used in the rain-runoff process includes the convolutional Artificial Neural Networks (ANN) models that are used to predict rainfall [1,2], estimate evaporation [3], predict plant water uptake [4], classify plants based on leaves [5], and determine crop yield [6]. The biological process and organization's structure employed the visual data from the UAV.

Figure 3.2 shows the drones' value with respect to application. Multi-spectral imagery can be recorded, and images can be captured at varied resolutions, depending on the height at which the operator is flying. Without machine learning methods, it has been challenging to understand high-resolution photographs [7]. Random forest (RF) is a hierarchical machine learning technique that uses bootstrapping, or aggregation, to classify images [8]. Estimates for spectral bands can be produced using the RF approach [9]. Deep learning methods and neural networks have less performance than random-forest-based modeling [10]. In quantitative satellite data, extreme learning machine methods are employed for classification and regression problems [11,12].

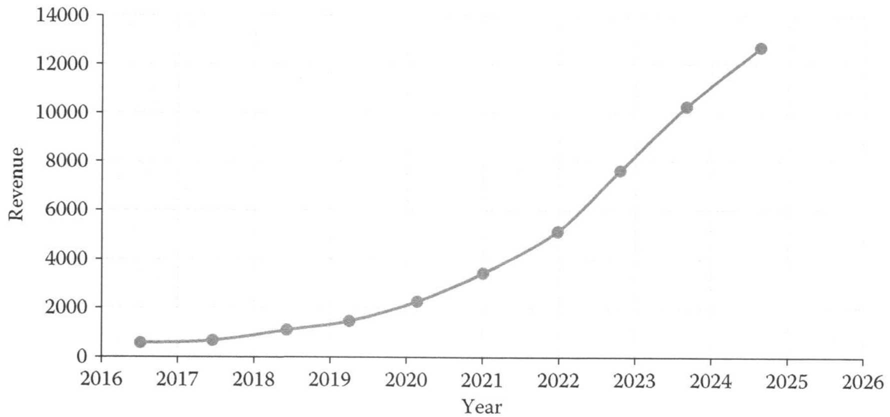

FIGURE 3.1 Revenue of drones during the period of 2016 to 2025.

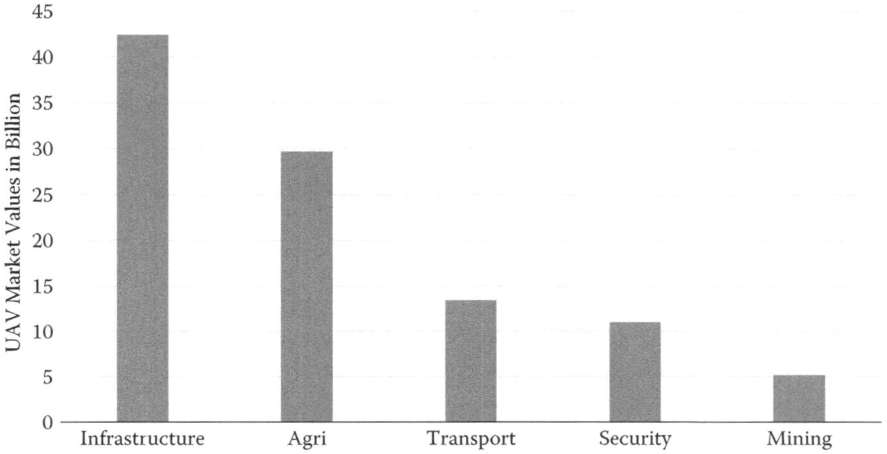

FIGURE 3.2 Drones value in with respect to application.

3.3 DRONE ECOSYSTEM IN THE MACHINE LEARNING INDUSTRY

The growing use of unmanned aerial vehicles (UAVs) in multiple industries has been made possible due to recent advances in machine learning techniques, sensors, and IT technologies. Despite the growing relevance of data communication and electronics, the primary industries include computers, wireless networks, smart cities, defense, farming, and mining.

3.3.1 Computers and Wireless Networks

A UAV has been integrated into a wireless network via several research projects. UAV-based networks employ deep convolutional neural networks to better understand the behaviors of the unmanned aerial vehicles (UAVs) under various environmental conditions, especially under remote control. Liquid state machine (LSM) is used as a machine learning model to help manage resources, as it is known to have greater precision in decision making and performance over other learning algorithms despite data variance [13–15].

3.3.2 Sustainable Smart Cities and Military Facilities

There has been an increase in the deployment of UAVs in smart urban cities and in the military for varied objectives. A graffiti removal system was developed using the UAV system and machine techniques and algorithms [16]. Using machine learning, a system was created that was able to distinguish between piloted and unpiloted flying objects. UAVs are found by tracking their controller signals. This is done by sending out RF signals from the controller [17]. It is determined how different UAV kinds were by utilizing the Wi-Fi signals detected between a UAV and its controller in combination with Markov, CNN, and Naive-Bayes classifiers [18].

3.3.3 Farming

Drones have proven their use in the prediction of irrigation water requirements and management of water supplies. This has been possible with the aid of deep learning and drone photography. The models also include Wild Jungle, Logistic Regression, and Multiple Regression for predicting and classifying vegetation production and crops [19–23].

3.3.4 Others

Some reports suggest the following kinds of information are gathered: statistics for wildlife, geology, and mining. UAV-enabled machine learning was utilized to detect animals. Using machine-learning methods, [24] the classification of geotechnical mining models was done with a classification algorithm, k-nearest neighbor, randomized forest, gradient tree booster, and classification relevance vector machine, which use surface feature detection. Table 3.1 contains a compilation of prior research about UAVs and machine learning.

Image categorization and object detection will benefit from deep learning and UAV mixing thanks to greater precision, accuracy, and more efficient outcomes. In the quadcopter and advanced algorithms research, the United States is dominant. A total of 38.9% of the study is focused on it, with the Asia Pacific region coming in next (China, Japan, and Korea). Even in 2017, Africa and Latin America are trailing in terms of drone and machine learning research, and those are the continents with the least amount of such projects. Japan, Korea, and Europe have had very little in the way of robotics innovation, whereas countries in the Asia Pacific region, like Japan and Korea, have pioneered the field. While their application is likely to grow, the applications for UAVs include photography, surveying, mapping, precision agriculture, sensing, and product delivery [29–31]. Image 3.3a and 3.3b represent the machine learning techniques for UAV-based communications.

TABLE 3.1
Summary of Drones and Machine Learning in Various Applications

Focus area and ML Algorithms	Insight
K-Nearest Neighbors in the Military, Random Forests, and Support Vector Machines [25]	Data Transfer and Surveillance.
Multiclass Relevance Vector Machine [6], Wild Life Support Vector Machine [7]	Animal activities and their behaviors studies.
Random Forest and Extreme Learning Machine in Agriculture [26,27]	Survey and monitoring of crops.
Detection and identification of UAVs Markov and Naive Bayes models [17]	Prediction on various levels.
Convolutional Neural Network for Traffic Management [28]	Safety and surveillance.
Support Vector Machine in Geology/Mining, k-Nearest Neighbors, Random Forest, Gradient Tree Boost [24]	Forest fires.
Machine Learning Algorithms for Classification [18]	Various applications.

(a)

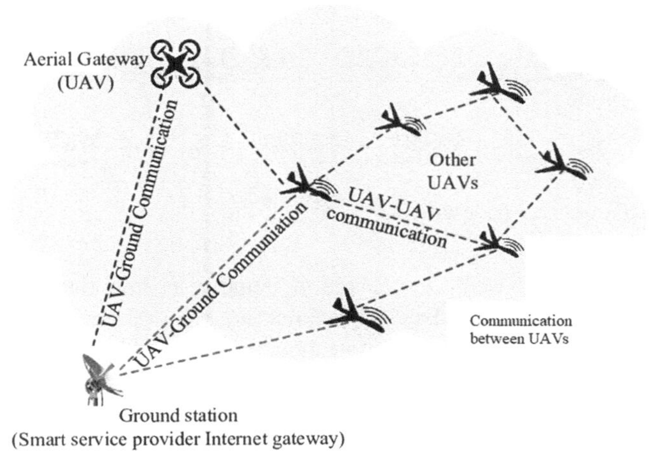

(b)

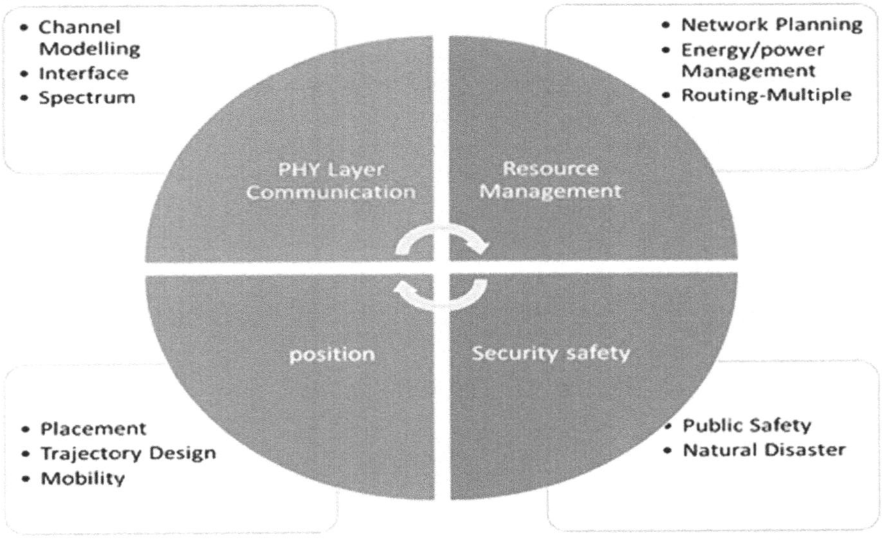

FIGURE 3.3 (a) UAV-based communications [32]; (b) classification of the AI/ML solutions in UAV-based [32].

To identify weed patches, UAVs are ideal products. The advantage of UAVs when compared to UTVs is shorter timing of monitoring or surveying time to control the presence of foreign materials [33]. In a shorter time, UAVs can take photographs and cover the agriculture land. UAVs are used with deep learning and other technologies for the use of agriculture and environment conditions [32,34–48]. The classification of drones can be conducted by the radar system with the help of machine learning.

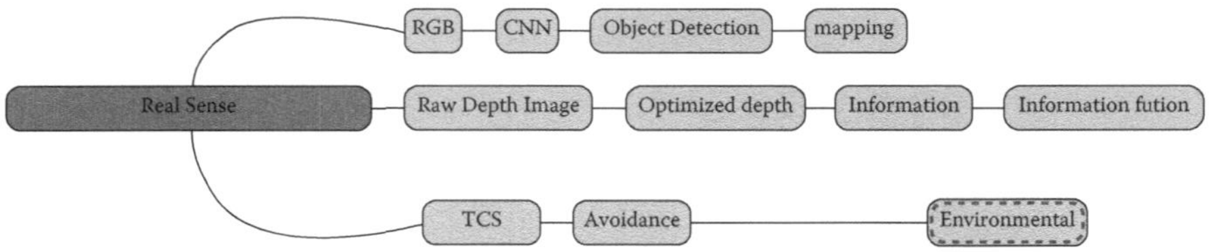

FIGURE 3.4 Flowchart ML in drones.

Drone classification based on radar return signal is an important task for public safety applications. Figure 3.4 depicts the flowchart ML in drones. Determining the make or class of a drone gives information about the potential intent of the UAV [49]. Table 3.2 contains different techniques of machine learning algorithms used in UAVs.

The drones can also be used to detect poachers and factors involved in machine learning were studied. The studies are focused on identifying the factors and efficiency of the ML approach [50]. The below image represents the autonomous obstacle avoidance using UAV [33]. In the last two decades, the application of drones has been used in medical applications [30]. It involves safety in flying, transmission of data, and privacy data. Several literatures are involved in the network and blockchain technology in the drone areas for medical applications [43,51].

In recent times, air taxiing is the robust transportation that is focused for the future. Figure 3.5 illustrates the emergency system that has been used in UAVs. The ML can be used to study weather conditions and demand for air taxis in aviation services [39]. The drones can be utilized in remote-sensing devices integrated with machine learning. They can be used in studies of wind erosion and dry lands for remote sensing applications [40].

TABLE 3.2
Algorithm of Drones and Machine Learning

Security Target	ML Solution
Eavesdropping mitigation	Q-learning
Interference and jamming mitigation	PHC-based learning
Jamming, spoofing, and eavesdropping mitigation	RL
GPS spoofing protection	ANN-supervised learning
Eavesdropping detection	One-class SVM and K-means
Interception of malicious UAVs	Q-learning
Real-time mapping	Genetic algorithm-based
UAV pilot identification	Classification based on LD, QD, SVM, KNN, or R and F
Guaranteeing privacy, integrity, and confidentiality of UAV compressed video streams	CNN-based detection
Protection against trespassing UAVs	SVM-based ML
Small boat detection	DL-based CNN

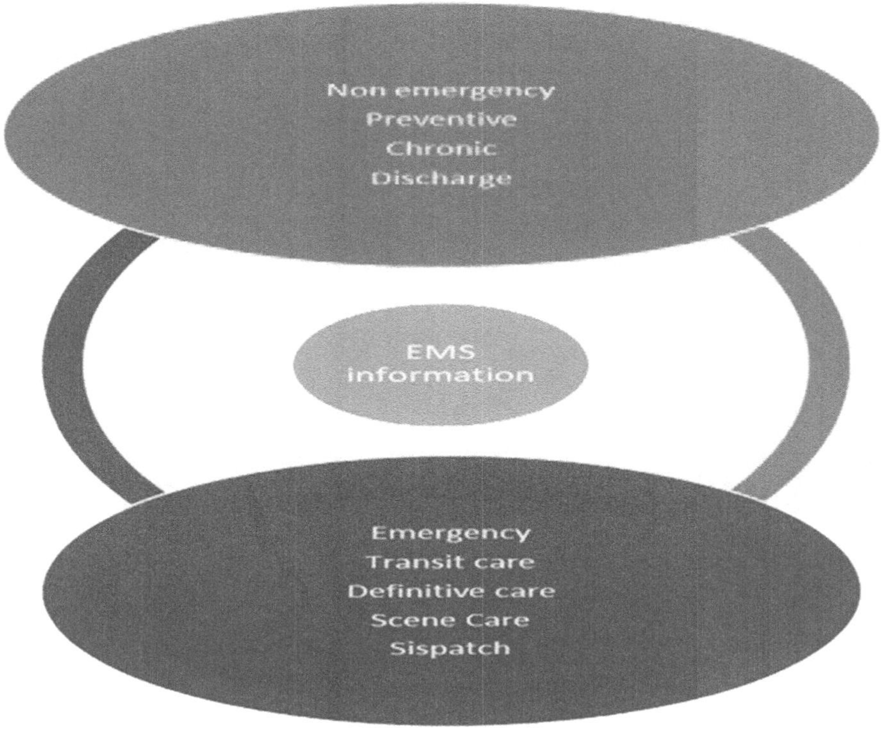

FIGURE 3.5 Emergency system [46].

3.4 MARKET ANALYSIS

See the projected market shares of UAVs around the globe by 2025 in Figure 3.6. The United States will have an advantage over other countries in 2025, thanks to increased deployment of UAVs in different areas. Asia Pacific will remain the market share leader for UAVs. Other countries that can be found here are Europe (9% share) and the Middle East (8% share). Research into drones and machine learning in the service of smart cities and the military has really taken off since 2016. While only 20% of marketers were doing it in 2016, it had increased to 40% in 2019. There is steady progress in research on UAVs and machine learning in the agricultural sector.

Figure 3.7 illustrates that the diverse types of machine learning algorithms shared space in various industries during the past four years. Many algorithms were employed in the relationship of UAVs with machine learning. The random forest has the highest share of algorithms among all. The flexibility to manage noise in the data is what has made it the most often used algorithm. Twenty-one percent of support vector machine users use it as their second choice. It is notable that the use of 16% deep convolutional neural networks and 11% K-nearest neighbors

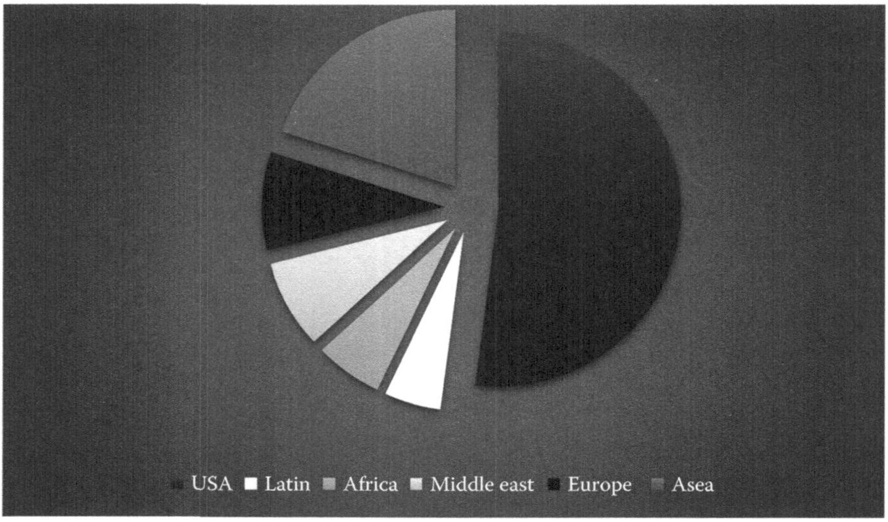

FIGURE 3.6 Drone project by global region in 2025.

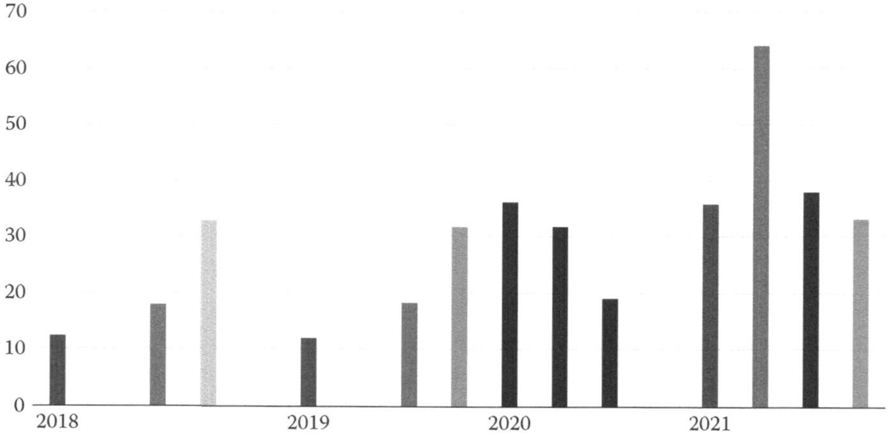

FIGURE 3.7 Usage of ML algorithms in Drones from the year 2018 to 2021.

implementation grew dramatically. The additional algorithms that have been utilized only on occasion are liquid state, multi-agent learning, artificial neural network, and (among others) liquid state.

UAV research has been employing machine learning algorithms, shown in Figure 3.8, over the past four years. Although certain difficulties may be linked to UAVs and machine learning utilization, there are still issues. UAV-assisted wireless networks worry drone manufacturers. Operators and employers are often the ones who suffer the most when they deal with the cognitive and manpower burden

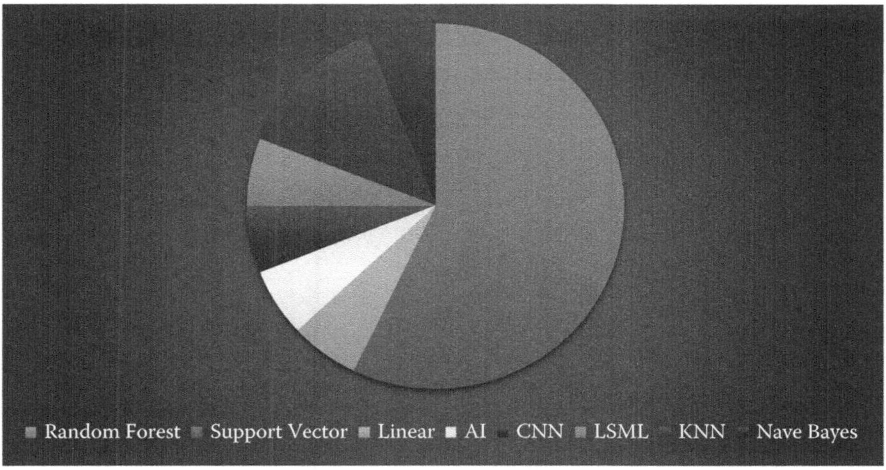

FIGURE 3.8 ML procedures employed in drone research.

caused by the adoption of UAVs. More people are needed to control the UAVs than to control other aircraft. As privacy invasion tools, they use UAVs with facial recognition software and infrared technologies. Sometimes consumer drones evade radar, vision, and sound technology when used. Civil societies have discussed the use of drones for civilian surveillance. Because enormous training data sets can be expensive and time consuming to work with, the machine learning field must come up with ways to deal with this issue.

Unmanned aerial robotic system component classification:

a. Hardware interfaces
b. Motor system
c. Feature extraction system
d. Situational awareness system
e. Executive system
f. Planning system
g. Supervision system
h. Communication system

3.4.1 Machine Learning–Assisted Framework

Table 3.3 summarizes the machine learning techniques with opportunities for the unmanned aerial vehicle applications.

3.4.2 Learning Models

In support of autonomous decision making, three specific deep learning models emerge most frequently in the research pool. To begin with, "VGG-16" [41] is a

TABLE 3.3
ML Opportunities and Techniques with Respect to its Functions

Functions	ML opportunities	ML techniques
Operating Planning	• Dynamic trajectory planning in new situations. • Lack of prior information of paths, needed to be classified. • Adjustment of paths as a function of UAV states.	• Reinforcement learning when no a priori data is available or the environment changes. • Deep reinforcement learning processes many parameters.
Situational awareness	• Weather condition awareness and prediction. • Terrain awareness and assessment of communication and safety precautions. • Obstacle awareness.	• Convolutional neural network to estimate the 3D wind flow for trajectory adjustment. • Support vector machine conditions of the UAV along the flight.
Failure Detection and Recovery	• Failure detection (UAV, network, ground users). • Failure assessments and impacts classifications. • Failure recovery.	• Supervised learning that uses historical databases for prediction of future failure. • Artificial neural network for risk assessment in the UAS network.
Remote Identification	• Data security and analysis.	• Data mining techniques to provide a secure analytic of dynamic streams of data.

CNN image classifier that was trained on the "ImageNet" data set [42], which has over 14 million pictures and thousands of labels matched. VGG-16 may be used to classify a wide range of photos or as a foundation for transfer learning with fine-tuning utilizing images particular to a target drone environment. The bulk of research works in the research pool that utilize it or the object detection model "YoloV3" [43] as a basis for collision avoidance or object detection/distinction use it as a base. The "ResNet" design [44] is based on a CNN-based article that discusses how residual layer "shortcuts" might imitate the activity of complete neural layers to improve the "AlexNet" architecture [45]. ResNet, like VGG-16, is trained on the ImageNet data set. The advantage of ResNet's shortcuts design is that it reduces processing overhead significantly, yielding in efficient models with fast reaction times and equivalent accuracy. This is advantageous for drone activities that need little CPU use. "DroNet" is more focused on autonomous drone navigation, and it uses manually labeled automobile and bicycle footage as training data for navigation in a city setting. DroNet outputs from a single image are particular to drone navigation, including a steering angle to keep the drone moving while avoiding objects, as well as a collision probability to allow the UAV to notice risky situations and respond quickly.

3.4.3 Challenges in the Current Research Work

Most research papers describe their strategy to model training and testing, including the ground truth they used, labels, and explanations of how the navigation system interacts with the CNN model. However, there is a lack of consistency in measures throughout the industry. The web address some articles assess their technique using measures that are particular to the context, such as performance at various speeds [20] and the number of successful laps [47]. In DNN (domain name system) there is the inclusion of visual descriptions of structures and evaluation findings in the research space. Comparing similar architectural or functional approaches is critical to the project's ability. When displayed without context, the use of research work-specific metrics is tough to compare the two because of the relationship to a more popular statistic like accuracy. Performance of autonomous navigation techniques in different environments across the domain. Another common difficulty in the research pool is numerous computer and electrical engineering roadblocks that have not been attempted to be overcome or addressed, or solutions that have been carefully constructed to function within the confines of such roadblocks. This affects the implementation's resilience and may limit the use cases in which the solution may be used.

This problem affects power consumption, data processing, delay, sensor design, and communication. We believe that multidisciplinary collaboration might be extremely beneficial to drone autonomy research.

3.5 CONCLUSION

The UAV has a lot of good qualities, such as flying at a low height, being compact, having excellent resolution, being lightweight, and being portable. The benefits of the UAV and machine learning are particularly large in scientific research. The research analyzed here explored the uses of UAVs and machine learning applications. The research in combined UAV and machine learning has not matured yet. The study's findings show that a lot of the studies in this field are scattered, and the research mostly focuses on computer and wireless network connections, smart cities, military work, agriculture, mining, and data analysis on wildlife. The UAV relies heavily on the support vector and random forest algorithms. The USA and Asia Pacific region play a huge role in the amount of UAV research and implementation. The presence of a large number of commercial UAVs operating without registration has generated security and privacy concerns. There is speculation about the construction of a model that can recognize unregistered consumer UAV in the future. Advanced models trained to recognize items will be built for the future so that unmanned aerial vehicles and satellite imagery may be parsed to extract diverse types of objects.

ACKNOWLEDGMENT

This research received funding from The Institution of Engineers (India) Ref No. R.4/2/UG/2017-18/RDUG2017026 and Tamilnadu State Council for Science and Technology Ref No. TNSCST/DIT/03/VR/2017-2018.

REFERENCES

[1] Wagner, I., Projected commercial drone revenue worldwide 2016–2025. 2019. Statista GmbH, J.B.Platz 1 20355 Hamburg, Germany.
[2] Sudheer, K.P., A.K. Gosain, D. Mohana Rangan, and S.M. Saheb, Modelling evaporation using an artificial neural network algorithm. *Hydrological Processes*, 2002. 16(16): pp. 3189–3202.
[3] Qiao, D.M., H.B. Shi, H.B. Pang, X.B. Qi, and F. Plauborg, Estimating plant root water uptake using a neural network approach. *Agricultural water management*, 2010. 98(2): pp. 251–260.
[4] Wu, Stephen Gang, Forrest Sheng Bao, Eric You Xu, Yu-Xuan Wang, Yi-Fan Chang, and Qiao-Liang Xiang. A leaf recognition algorithm for plant classification using probabilistic neural network. in 2007 IEEE international symposium on signal processing and information technology. 2007: IEEE.
[5] Khairunniza-Bejo, Siti, Samihah Mustaffha, and Wan Ishak Wan Ismail, Application of artificial neural network in predicting crop yield: A review. *Journal of Food Science and Engineering*, 2014. 4(1): p. 1.
[6] Zaman, Bushra, Dr Mac Mckee, and Austin Jensen, UAV, Machine Learning, And GIS for Wetland Mitigation in Southwestern Utah, USA. 2017.
[7] Rey, Nicolas, Combining UAV-imagery and machine learning for wildlife conservation. 2016.
[8] Gislason, Pall Oskar, Jon Atli Benediktsson, and Johannes R. Sveinsson, Random forests for land cover classification. *Pattern Recognition Letters*, 2006. 27(4): pp. 294–300.
[9] Douglas, Reward K., Said Nawar, M. Carmen Alamar, A.M. Mouazen, and Frederic Coulon, Rapid prediction of total petroleum hydrocarbons concentration in contaminated soil using vis-NIR spectroscopy and regression techniques. *Science of the Total Environment*, 2018. 616: pp. 147–155.
[10] Peng, Jie, Asim Biswas, Qingsong Jiang, Ruiying Zhao, Jie Hu, Bifeng Hu, and Zhou Shi, Estimating soil salinity from remote sensing and terrain data in southern Xinjiang Province, China. *Geoderma*, 2019. 337: pp. 1309–1319.
[11] Maimaitijiang, Maitiniyazi, Abduwasit Ghulam, Paheding Sidike, Sean Hartling, Matthew Maimaitiyiming, Kyle Peterson, Ethan Shavers, Jack Fishman, Jim Peterson, and Suhas Kadam, Unmanned aerial system (UAS)-based phenotyping of soybean using multi-sensor data fusion and extreme learning machine. *ISPRS Journal of Photogrammetry and Remote Sensing*, 2017. 134: pp. 43–58.
[12] Khosravi, Vahid, Faramarz Doulati Ardejani, Saeed Yousefi, and Ahmad Aryafar, Monitoring soil lead and zinc contents via combination of spectroscopy with extreme learning machine and other data mining methods. *Geoderma*, 2018. 318: pp. 29–41.
[13] Chen, Mingzhe, Walid Saad, and Changchuan Yin, Liquid state machine learning for resource and cache management in LTE-U unmanned aerial vehicle (UAV) networks. *IEEE Transactions on Wireless Communications*, 18(3): pp. 1504–1517, 2019.
[14] Mandloi, Yograj S. and Yoshinobu Inada. Machine Learning Approach for Drone Perception and Control. in International Conference on Engineering Applications of Neural Networks. 2019: Springer.
[15] Liu, Xiao, Yuanwei Liu, Yue Chen, and Lajos Hanzo, Trajectory design and power control for multi-UAV assisted wireless networks: a machine learning approach. arXiv preprint arXiv:1812.07665, 2018.

[16] Nahar, Prakhar, Kang-hua Wu, Sitao Mei, Hadiyah Ghoghari, Preethi Srinivasan, Yueh-lin Lee, Jerry Gao, and Xuan Guan. Autonomous UAV Forced Graffiti Detection and Removal System Based on Machine Learning. in 2017 IEEE SmartWorld, Ubiquitous Intelligence & Computing, Advanced & Trusted Computed, Scalable Computing & Communications, Cloud & Big Data Computing, Internet of People and Smart City Innovation (SmartWorld/SCALCOM/UIC/ATC/CBDCom/IOP/SCI). 2017: IEEE.

[17] Ezuma, Martins, Fatih Erden, Chethan Kumar Anjinappa, Ozgur Ozdemir, and Ismail Guvenc, Micro-UAV Detection and Classification from RF Fingerprints Using Machine Learning Techniques. arXiv preprint arXiv:1901.07703, 2019.

[18] Alipour-Fanid, Amir, Monireh Dabaghchian, Ning Wang, Pu Wang, Liang Zhao, and Kai Zeng, Machine Learning-Based Delay-Aware UAV Detection and Operation Mode Identification over Encrypted Wi-Fi Traffic. arXiv preprint arXiv:1905.06396, 2019.

[19] Kwak, Geun-Ho and No-Wook Park, Impact of texture information on crop classification with machine learning and UAV images. *Applied Sciences*, 2019. 9(4): p. 643.

[20] Viljanen, Niko, Eija Honkavaara, Roope Näsi, Teemu Hakala, Oiva Niemeläinen, and Jere Kaivosoja, A novel machine learning method for estimating biomass of grass swards using a photogrammetric canopy height model, images and vegetation indices captured by a drone. *Agriculture*, 2018. 8(5): p. 70.

[21] Jónsson, Sigurbjörn, RGB and Multispectral UAV image classification of agricultural fields using a machine learning algorithm. *Student thesis series INES*, 2019.

[22] Damgaard, Christian, Integrating hierarchical statistical models and machine learning algorithms for ground-truthing drone images of the vegetation: taxonomy, abundance and population ecological models. *bioRxiv*, 2018: p. 491381.

[23] Suzuki, T., T. Tsuchiya, S. Suzuki, and A. Yamamba, Vegetation classification using a small UAV based on superpixel segmentation and machine learning. *Journal of The Remote Sensing Society of Japan*, 2016. 36(2): pp. 59–71.

[24] Beretta, Filipe, A.L. Rodrigues, R.L. Peroni, and JFCL Costa, Automated lithological classification using UAV and machine learning on an open cast mine. *Applied Earth Science*, 128(3): pp. 79–88. 2019.

[25] Monfort, Samuel S., Ciara M. Sibley, and Joseph T. Coyne. Using machine learning and real-time workload assessment in a high-fidelity UAV simulation environment. in *Next-Generation Analyst IV*. 2016: International Society for Optics and Photonics.

[26] Ge, Xiangyu, Jingzhe Wang, Jianli Ding, Xiaoyi Cao, Zipeng Zhang, Jie Liu, and Xiaohang Li, Combining UAV-based hyperspectral imagery and machine learning algorithms for soil moisture content monitoring. *PeerJ*, 2019. 7: p. e6926.

[27] Romero, Maria, Yuchen Luo, Baofeng Su, and Sigfredo Fuentes, Vineyard water status estimation using multispectral imagery from an UAV platform and machine learning algorithms for irrigation scheduling management. *Computers and Electronics in Agriculture*, 2018. 147: pp. 109–117.

[28] Siewert, Sam, Mehran Andalibi, Stephen Bruder, Iacopo Gentilini, Aasheesh Dandupally, Soumyatha Gavvala, Omkar Prabhu, Jonathan Buchholz, and Dakota Burklund, *Drone Net, a passive instrument network driven by machine vision and machine learning to automate UAS traffic management*. 2018, AUVSI Xponential, Denver, Colorado.

[29] Hassanein, M. and N. El-Sheimy An efficient weed detection procedure using low-cost UAV imagery system for precision agriculture applications. *International Archives of the Photogrammetry, Remote Sensing and Spatial Information Sciences*. 2018. 42: pp. 181–187.

[30] Prasad G., Abhishek P., and Karthick R., Influence of unmanned aerial vehicle in medical product transport. *International Journal of Intelligent Unmanned Systems*, 7(2): pp. 88–94, 2018. 10.1108/IJIUS-05-2018-0015
[31] Prasad, G., Performance estimation of twin propeller in unmanned aerial vehicle. *INCAS Bulletin*, 12(2):2018 pp. 143–149. 10.13111/2066-8201.2020.12.2.12
[32] Alshallash, K.S., Germination of weed species (Avena fatua, Bromus catharticus, Chenopodium album and Phalaris minor) with implications for their dispersal and control. *Ann Agric Sci*. 2018. 63: pp. 91–97.
[33] Prasad G., Vijayaganth V., Sivaraj G., Rajasekar K., Ramesh M., Gokul raj R., and Matheeswaran P., Positioning of UAV using Algorithm for monitoring the forest region. *IEEE Xplore Digital Library*, pp. 1361–1363, 2018. 10.1109/ICISC.2018.83 99030
[34] Naveen Kumar K., Prasad G., Rajasekar K., Vadivelu P., Satyanarayana Gollakota, and Kavinprabhu S.K., A study on the forest fire detection using unmanned aerial vehicles. *International Journal of Mechanical and Production Engineering Research and Development (IJMPERD)*, 2018. 8(7): pp. 165–171.
[35] Raja Sekar, K., D. Vignesh Moorthy, G. Prasad, P. Manigandan, M. Senthil Kumar, An experimental investigation on the influence of bio-inspired tubercle in the unmanned aerial vehicle propeller. *International Journal of Mechanical and Production Engineering Research and Development (IJMPERD)*, 2018. 8(7): pp. 404–409.
[36] Krishna, K.R. *Agricultural drones: a peaceful pursuit*. Taylor Francis. 2018.
[37] Crimaldi, M., V. Cristiano, A. De Vivo, M. Isernia, P. Ivanov, and F. Sarghini Neural network algorithms for real time plant diseases detection using UAVs. In: Coppola A., Di Renzo G.C., Altieri G., D'Antonio P., editors. *Innov Biosyst Eng Sustain Agric For Food Prod*. 2020. pp. 827–835.
[38] Fawakherji, M., C. Potena, D.D. Bloisi, M. Imperoli, A. Pretto, and D. Nardi, *UAV Image based crop and weed distribution estimation on embedded GPU boards BT computer analysis of images and patterns*. Berlin: Springer; 2019.
[39] Rasmussen, J., J. Nielsen, J.C. Streibig, J.E. Jensen, K.S. Pedersen, and S.I. Olsen, Pre-harvest weed mapping of Cirsium arvense in wheat and barley with off-the-shelf UAVs. *Precision Agriculture*, 2019. 20: pp. 983–999.
[40] Franco, C., C. Guada, J.T. Rodríguez, J. Nielsen, J. Rasmussen, D. Gómez, et al. Automatic detection of thistle-weeds in cereal crops from aerial RGB images BT information processing and management of uncertainty in knowledge-based systems. *Applications*, 2018. pp. 441–452.
[41] Louargant M., G. Jones, R. Faroux, J.N. Paoli, T. Maillot, C. Gée, et al. Unsupervised classification algorithm for early weed detection in row-crops by combining spatial and spectral information. *Remote Sensing*, 2018. 10: pp. 1–18.
[42] Uang, Y., K.N. Reddy, R.S. Fletcher, and D. Pennington, UAV low-altitude remote sensing for precision weed management. *Weed Technology*, 2018. 32: pp. 2–6.
[43] Hansen, K.D., F. Garcia-Ruiz, W. Kazmi, M. Bisgaard, A. La Cour-Harbo, J. Rasmussen, et al. An autonomous robotic system for mapping weeds in fields. *IFAC Proc Vol IFAC-Pap*, 2013. 8: pp. 217–224.
[44] Louargant, M., S. Villette, G. Jones, N. Vigneau, J.N. Paoli, and C. Gée, Weed detection by UAV: simulation of the impact of spectral mixing in multispectral images. *Precision Agriculture*, 2017. pp. 1–20.
[45] López-Granados, F., M. Jurado-Expósito, J.M. Peña-Barragán, and L. García-Torres, Using remote sensing for identification of late-season grass weed patches in wheat. *Weed Science*, 2006. 54: pp. 346–353.
[46] Sanders, J.T., W.J. Everman, R. Austin, G.T. Roberson, and R.J. Richardson, Weed species differentiation using spectral reflectance land image classification. *ProcSPIE*, 11007: pp. 109–117, 2019.

[47] Davis S., Mangold J., Menalled F., Orloff N., Miller Z., and Lehnhoff E. A Meta-analysis of Canada Thistle (Cirsium arvense) management. *Weed Sci*ence, 2018. 66: pp. 548–557.

[48] De Castro, A.I., J. Torres-Sánchez, J.M. Peña, F.M. Jiménez-Brenes, O. Csillik, and F. López-Granados, An automatic random forest-OBIA algorithm for early weed mapping between and within crop rows using UAV imagery. *Remote Sensing*, 2018. 10: p. 285.

[49] Hudson, Sinclair, and Bhashyam Balaji. Application of machine learning for drone classification using radars. In *Signal Processing, Sensor/Information Fusion, and Target Recognition XXX*, vol. 11756, p. 117560C. International Society for Optics and Photonics, 2021.

[50] Doull, Katie E., Carl Chalmers, Paul Fergus, Steve Longmore, Alex K. Piel, and Serge A. Wich. An Evaluation of the Factors Affecting 'Poacher'Detection with Drones and the Efficacy of Machine-Learning for Detection. *Sensors*, 2021. 21(12): 4074.

[51] Kavya, J., G. Prasad, and N. Bharanidharan. Artificial Intelligence, Machine Learning, and Internet of Drones in Medical Applications. *Bio-Inspired Algorithms and Devices for Treatment of Cognitive Diseases Using Future Technologies. IGI Global*, 2022. 180–188.

4 Review of Medical Drones in Healthcare Applications

Kavitha Rajamohan

CONTENTS

DOI: 10.1201/9781003252085-4

4.1 INTRODUCTION

The drone, also known as the unmanned aerial vehicle (UAV), is an unmanned aircraft without a pilot that is activated by many technologies such as computer vision, artificial intelligence (AI), Internet of Things (IoT), cloud computing, and big data along with cutting-edge security techniques. Drone technology is the new major growing invention in the transportation field. Drones are also called remotely piloted aircraft (RPA), remotely operated aircraft (ROA), unmanned aircraft (UA), and remotely piloted vehicle (RPV). There is an extensive range of probable drone applications, such as exploration of agricultural crops, monitoring weather conditions, relay communication and broadcasting, inspection of infrastructure, study of the damage level during disasters, search and rescue, understanding the traffic flow, automated security, on-demand parcel delivery, wildlife surveillance, cinematography, and so on. This chapter discusses the architecture and applications of drones in health care.

4.1.1 IoD–Internet of Drones

Understanding of engineering principles involved in drones is important. The forces the drone experiences during flight are thrust, drag, weight, and lift. Thrust helps the drone move in the direction of motion, drag slows down the drone in flight, weight is caused by gravity, and lift holds the drone in the air. There are three types of acceleration experienced by drones at the time of flight-linear acceleration, which reflects a change in speed in a straight line that occurs the time of takeoff and landing; radial acceleration, which happens during sharp turns; and angular acceleration which comes into play during spins and climbing turns [1]. Generally, there are two types of drones: (i) fixed wing, which makes it possible for a lift to sustain velocity through the air as it moves, and (ii) the rotor, which is highly maneuverable and can hover and rotate with a flight controller. With respect to application, drones are chosen based on required payload, flight range, and battery capacity. Recently, a steep growth revolved around the advancement of drone hardware, software, and networks.

The layered architecture particularly designed to coordinate drones is called Internet of Drones (IoD). This architecture is designed to provide navigation services from one location to another, which are referred to as nodes in the network. The proposed [2] conceptual framework of IoD architecture consists of three large-scale networks, namely the Internet, the cellular network, and air traffic control network.

4.1.2 Drones in Health Care

Health care took a great leap forward in the past decade. In recent years, the healthcare industry has been focusing on providing the best services using technological advancement integrated with existing medical services. The availability of the Internet and advancement in medical devices has made it possible to provide healthcare services anytime and anywhere. The development in telemedicine is a great treat to the patients when they must avoid visiting hospitals, especially during the COVID-19 pandemic. Nanomedicine, a medical application of nanotechnology, created a path to deliver drugs to tumours using gold nanoparticles. The combination of AI, IoT, and big data with 5G has made the healthcare system provide predictive, preventive, personalized, and anticipatory service in real time. Many wearable devices are introduced with a purpose, like monitoring glucose and blood sugar levels. Sometimes technological development may not reach the needy due to unexpected natural disasters. In this situation, drone technology supports healthcare providers and technology reaches the needy. Healthcare experts believe that drones can deliver blood samples, vaccines, and organs faster and safer to save millions of lives.

Drones play an important role in civil emergency medical care. A drone or a swarm of drones connected via a network that is built for specific healthcare applications is called Internet of Medical Drone (IoMD). The usage of drones for medical purposes has many benefits, such as quicker help, shortening the traveling time to reach patients, supporting medical agency teams, and transporting medicine to unreachable places. Drones are commonly used in COVID-19 pandemic emergency situations [3] to minimize the risk and spread of infection among healthcare professionals who are prone to coronavirus infection. Drone-aided shipments and collection of medicine and test kits is introduced for chronic patients to avoid regular visits to the clinic for their health check-ups.

4.2 INTERNET OF MEDICAL DRONE (IOMD) IN HEALTH CARE

Drone technology affords numerous advantages and opportunities in a wide range of domains. Nowadays, the application of IoMDs in health care is increasing. Initially, drones were developed only for military operations. Later, because of their efficient and enhanced level of safety and security, they found their way to support the public. Drones are used to deliver medicines and test kits, to monitor COVID pandemic situations, etc., in remote area where the transportation is challenging. Worldwide, many countries such as China, United States, Austria, and Sweden demonstrated the usage of drones in different way. The specific survey and summary of several studies that are discussed in this section are given in Table 4.1.

4.3 ARCHITECTURE OF HEALTHCARE IOMD

Nowadays, drones are used in healthcare applications, especially during periods when healthcare providers are not able to access the patients in-person. The architecture of IoMD consists of three phases: (i) Data collection phase: Drones

TABLE 4.1
IoMDs in Health Care

Article Title	Year	Country of Study	Application-Focused	Targeted Population
Francher et al. [1]	2017	China	Delivery of medical supply	Remote areas of China
Kim et al. [4]	2017	United States (US)	Shipment and collection of medical supply and test kits	Rural areas
Ling et al. [5]	2019	African countries	Aerial delivery of medical supplies	Remote communities
Claesson [6]	2016	Sweden	Transporting an automated external defibrillator for out of hospital care	Patients
Li et al. [7]	2021	United States (US)	Treatment management system	Elderly people at senior care centres.
Kumar et al. [8]	2021	–	COVID-19 healthcare protection system	COVID-19 affected area
Kunovjanek and Wankumuller [9]	2021	Austria	Delivery of test kit to infected patients	COVID-19 patients
Poljak and Sterbenc [10]	2016	Tanzania	Transportation	Drone delivery to save life
Ghelichi et al. [11]	2017	Louisville, KY	Delivery of medicali tems	People from hard-to-access locations
Wankmuller et al. [12]	2021	Australia and Italy	Rescue missions	People living in mountain regions

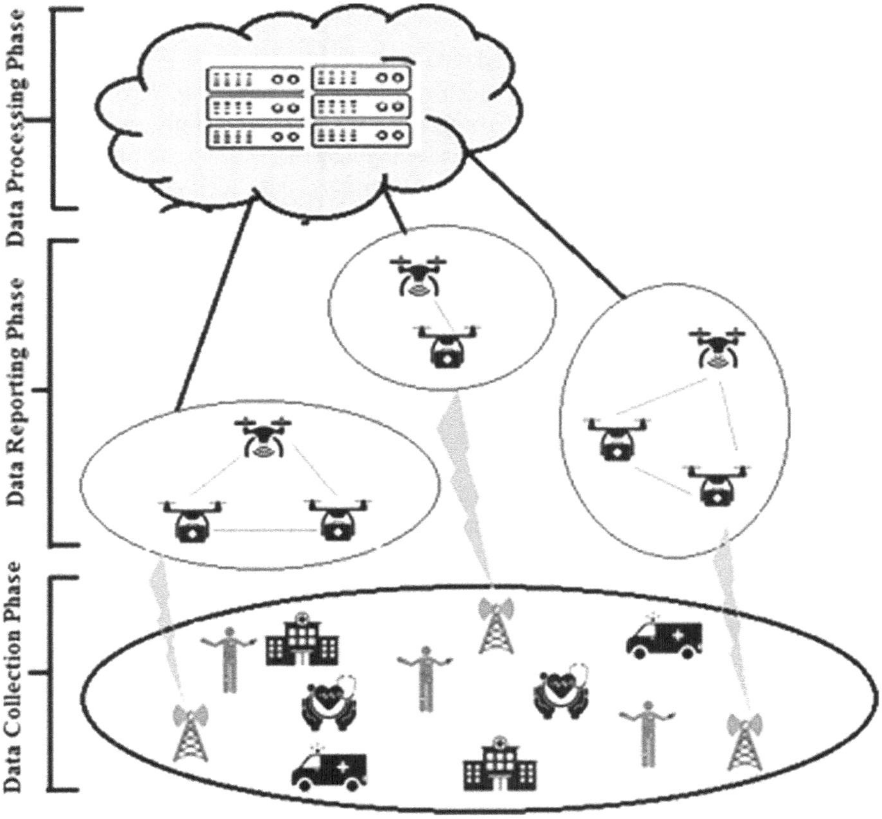

FIGURE 4.1 Architecture of IoMD.

collect patient information from various sources such as body area network, ambulance, etc.; (ii) Data reporting phase: Drones send the collected data to cloud server, medical server etc.; (iii) Data processing phase: Servers process the received data and help healthcare professionals make health decisions in real time Figure 4.1.

4.3.1 Data Collection Phase

Healthcare systems are impeccably integrated with different technologies, such as micro- and nano-technologies, wearable computing, and pervasive computing to facilitate healthcare systems for needy patients. Data from different sources like healthcare centers, patients' wearable body area network (BAN), and ambulances and Internet of Vehicles (IoV) are collected in this phase to enable healthcare services in the form of remote health monitoring. Sensed patients' data is received by drones via different communication networks.

4.3.2 Data Reporting Phase

In this phase, drones send the collected data to the application related servers. Drones are defined as small aerial vehicles which can flight autonomously, carry loads, and control remotely. Generally, drones have five basic units [13]. The first unit is a set of sensors like a barometer, a pilot tube, a current/voltage sensor, an inertial measuring unit, distance, temperature, and magnetometer sensors. These sensors are used for the operational purposes of drones such as measuring velocity, speed, pressure, temperature for stabilization, localization, and navigation. The second unit is the autopilot, which receives information from all sensors and keeps the drone in balanced flight. The third unit is the communication unit, which consists of a video controller, remote controller, and telemetry module. Drones can fly autonomously or by remote control pilot. As regulated by law, to ensure safety, drones are expected to be in remote control coverage areas. The fourth unit is the ground control station which streams real time data from the drone's camera. The final unit is the power source battery, which is lightweight and durable. For a successful application, all of these five components function together as well as connect with the health center. BAN and IoV are related examples to keep track of the health status of a panic attack and heart failure patients.

4.3.3 Data Processing Phase

Data sensed from the healthcare system is collected by the monitoring drone and transmitted to the data processing phase, which consists of technology infrastructures such as a cloud server and main server. For example, the drone that is implemented for patient health care keeps sensing data from BAN of heart patients and sends this data to cloud servers in real time. Healthcare providers can keep monitoring the BAN patients and provide quick, reliable access in life-critical situations whenever required.

4.3.4 Tools and Technologies of IoMD

The overall design of IoMD consists of a collection of hardware components and software tools. A handful of existing software tools, such as Paparazzi, APM, MultiWii Copter, KK, Djinaza, and pixhawk are available for drone prototyping. The hardware components are chosen suitable to the specific application by considering the weight, balance, and overall center of mass. Basically, the IoMDs consist of a flight control unit, a data transmission module, a data receiving module, and a video transmission module. These modules are designed with hardware components such as GPS and compass, remote controller, image transmitter, camera, gimbal, speed controller, motor, propeller, battery, motor power hub, servo(s), voltage sensor, current sensor, and remote controller [1].

4.4 APPLICATIONS OF IOMD HEALTHCARE SYSTEM

Generally, drones are classified into many categories based on size, performance, and application. Based on design, drones can be classified as (1) operational profiles, like horizontal or vertical take-off and landing; (2) regulatory type, like heavy or light; (3) governmental standard for specific operations; and (4) application-specific configuration, such as surveillance, intelligence, and investigation, payload delivery, and supply and communication network [14]. Applications of IoMD for the healthcare systems can be divided into three broader categories, such as search and rescue, transport and delivery, and medical care.

4.4.1 Search and Rescue

Drones have been used to detect health hazards, such as the levels of heavy metals like copper in farming land, aerosol (harmful gas) levels in the air, and radiation from uranium mines. Drones are used for observation of disaster location, places with biological and chemical exposure, and track virus spread. Drones are used to collect the number of patients who need help in high-risk situations. Drones can be used to assess initial damage and measure the health hazards related to chemical and biological materials, as evident in the Philippines after Typhoon Haiyan [14]. Usage of drones for healthcare services is a growing application area in many countries. Drones are a gift for healthcare providers to increase the ability and efficiency to provide their service to the person, particularly those in hard-to-reach locations. Different people have diverse health challenges, interests, policy priorities, and have connected to the IoMD for different health needs. Drones can save time and human resources by facilitating real-time aerial images of specific targeted areas, which helps search and rescue teams understand where help is very urgently needed. For example, a swarm of different drones can be implanted with different medical aid kits that can be distributed to a person who needs emergency care even before the rescue team reaches them, which can result in saving lives [15].

4.4.2 Transport and Delivery

Drones are used as a medical transport system to deliver microbiological samples more professionally. This started in 2007; researchers from the National Health Laboratory Service (NHLS) and the drone division jointly proved that drones are efficient to transport microbiological samples for human immunodeficiency virus (HIV) test. In 2014, researchers proved that drones can deliver laboratory samples by saving 75% of time than delivering samples by land for tuberculosis testing. Both studies proved that after delivery, there is no biological difference in samples delivered by drone and the traditional way [14]. Another study [15] proved that drones are capable of transporting diagnostic clinical laboratory specimens without affecting the nature of chemistry, hematology, and coagulation test results.

4.4.3 Medical Care

One of the emerging challenging applications of the drone is telemedicine, a remote treatment of patients using telecommunication technology. In 1998, a senior author from Greece demonstrated the establishment of instant telecommunication infrastructure (ITI) using drones to monitor pre- and post-operation of patients and provide remote guidance from experienced surgeons to juniors [14]. The author also mentioned that drones are used to provide wireless communication between the surgeon and robot to carry out telesurgery.

4.5 USAGE OF IOMD FOR HEALTH CARE

The usage of drone technology in healthcare applications is still in its infancy in most countries. A scoping review [16] determined that drones are used for healthcare services in the North American countries: Canada, the United States, and Mexico. Another study [5] was done to review the delivery and pickup of medication and test kits for chronic patients in rural areas of the USA. In Africa, drones are used to transport the blood samples of newborn babies to labs for HIV testing. In many countries, such as Taiwan and Nepal, drones are used to reach remote villages to deliver medical aids and collect test samples. In India, drones are used to handle disaster relief and rescue operations. During the Kumbh Mela festival, the Uttar Pradesh state police used drones to maintain law and order and safety of the devotees. In Uttarakhand, drones are used for successful delivery of blood samples from remote areas.

4.5.1 Drone Functions to Combat COVID-19 in India

During this pandemic, a non-government and non-profit organization, The Federation of Indian Chambers of Commerce & Industry (FICCI), recommended to the Indian government the usage of drone technology for solutions of many issues. The meeting with 40 stakeholders in attendance reviewed and understood the usage of drones that can play a momentous role by providing the solution to the problems caused by the COVID-19 pandemic [17]. As per the guidance given by this committee, drones are used in different Indian states for various purposes by the front-end warriors, namely police, healthcare, and municipal authorities. Figure 4.2 shows the usage of drones for COVID-19 impact management.

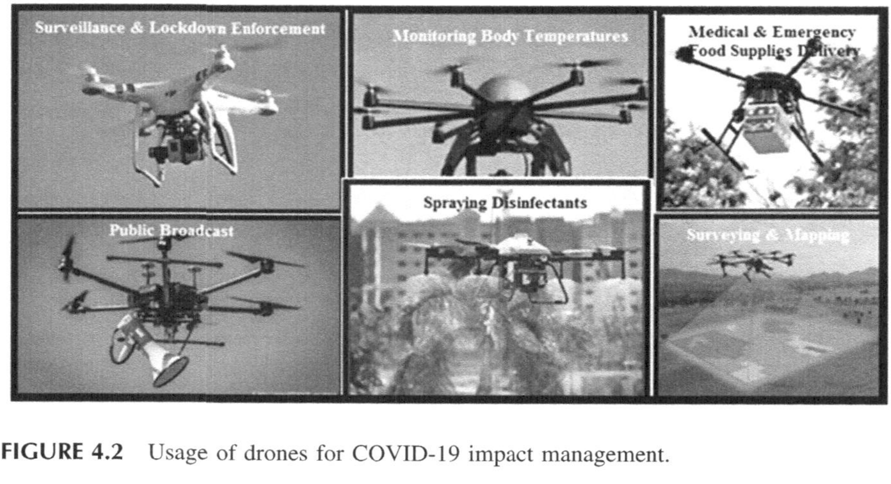

FIGURE 4.2 Usage of drones for COVID-19 impact management.

4.5.1.1 Surveillance and Lockdown Enforcement

The Indian police department adopted the drone technology to manage the COVID-19 pandemic situation. The police department of Lucknow used a weaponized drone which consists of pepper spray and a high-resolution camera for crowd control. In Maharashtra, one of the worst-hit states, Mumbai police used AI with drone technology to enforce the lockdown and social distancing on violators to control the spread in the state.

4.5.1.2 Public Broadcast

In India, in many states such as Kerala, Telangana, Jammu and Kashmir and Maharashtra, speaker mounted drones were used to spread awareness on COVID-19, to disperse crowds, and to encourage the public to stay at home. Especially in Kerala, siren-mounted drones were used to create awareness in public.

4.5.1.3 Monitoring Body Temperatures

Delhi municipal corporation launched a pilot project using drones with thermal scanners to spot people with higher temperatures. This project aims to catch temperature anomalies by scanning the large crowd by flying the thermal drone above the road and market. If the screened temperature is above the threshold, then the individual can be asked for further screening.

4.5.1.4 Medical and Emergency Food Supplies Delivery

North Delhi Municipal corporation used a multipurpose corona-combat drone which consists of an announcement speaker, thermal and night vision cameras, and a portable medical kit. A pilot project "Medicine from the Sky" was launched by the Telangana government in collaboration with Flipkart and Dunzo to supply vaccines and medicines to rural areas that have poor public transport infrastructure due to monsoons and other natural disasters. This project proved that drones can deliver vaccine supplies sooner and safer than by normal road transport.

4.5.1.5 Surveying and Mapping

Drones have played a vital role in survey mapping activity. Especially in hospital construction, and critical care facility arrangement, the drones were used to survey the area. During this pandemic, many countries like China, USA, and Germany have used drones for surveying the area with less human involvement to convert into hospitals.

4.5.1.6 Spraying Disinfectants

During COVID-19 pandemic, in the Telangana state, under Touch-me-Not project, a drone by the name Marut was used to spray mosquito insecticides in 2018 and used as a disinfectant sprayer. Marut can cover 20 km a day by carrying 10 liters of sanitizer while human sources can cover only 4 to 5 km a day.

4.6 RECENT TECHNOLOGIES IN HEALTHCARE IOMD

Software developers of healthcare drone designers always look for innovative technologies for a reasonable change. The technological revolution in drone design industry can be represented in six stages [18], as mentioned below:

Stage 1. All forms of basic remote control.
Stage 2. Static design with fixed camera and video recording, which was controlled by manual pilot.
Stage 3. Static design with two axis gimbals and HD video that is controlled by an assisted pilot with basic safety.
Stage 4. Transformative design with three axis gimbals with 1080P HD video which runs on autopilot with improved safety.
Stage 5. Transformative design with 360-degree gimbals with 4K video that is run by an intelligent pilot.
Stage 6. Drones that are suitable for commercial purposes, follow safety and regulatory standards, adaptable to platform and payload, air space aware, controlled by intelligent pilots and fully autonomous.

Nowadays, drone technologies are in their sixth stage. High-end professional companies have moved a little ahead in technologies by testing the drone with enhanced intelligent pilot mode with auto action such as takeoff, mission execution, and landing.

4.6.1 IoMDs

The hardware and software components of IoMDs are chosen based on the application design. General components used in drone design are discussed in an earlier section. Nowadays, the requirement makes the designing companies look for smart drones. To achieve this, smart sensors such as gyroscope, accelerometer, magnetometer, and GPS modules are used to control and monitor flight. Computer vision, object detection, and collision avoidance along with artificial intelligence will make drones smarter. Not only that, a built-in network enables coordination with other technologies to achieve the real-time application, such as patient monitoring. The drone self-monitoring systems are embedded to observe fuel level, damage of cameras, propellers, and sensors, and decide to take another route/action to avoid damage.

4.6.2 Cloud/Edge Computing

Along with drone technology, cloud/edge computing is used to analyze and make decisions based on the collected data. A multi-layered architecture is proposed [8] to collect and analyze data from patients with BAN. IoMDs are implemented to exchange data with edge, fog, and cloud servers for the essential computing, data exchange, and analytics. In this architecture, edge computing is used to avoid computational cost for local decisions. The cost of transferring data to a fog/cloud server is more in comparison with transferring to the edge. It also increases the scalability of devices connected in that layer. Edge computing helps to reduce the computational cost of entire data transmission and related decisions.

4.6.3 Internet Protocols

Mobile ad-hoc network, wireless sensor networks, body area network, and linear sensor networks are some commonly used networks to connect the drones to different healthcare applications. In order to increase output and efficiency, various network protocols such as the medium access control (MAC) protocol, MAVLink protocol, and ROSLink protocols were proposed by researchers. The MAVLink is a messaging protocol that is lightweight and ROSLink protocol communication protocol connects robots with IoT [19].

4.6.4 Artificial Intelligence

Nowadays, drone technologies are developed to become smarter with built-in capability of coordination with a network of drones and collaboration with other technologies for real time data analytics to make immediate decisions. AI is one such technology that is used in the delivery of defibrillators for out-of-hospital cardiac arrest [20]. In some healthcare applications, swarms of drones are used. In such cases, these drones are subject to security threats such as range of attacks or GPS jamming. A machine learning and software defined network is proposed [21] to secure the network communication in a swarm of drones.

4.6.5 Blockchain

With the recent advancement in communication and networking technologies, there is a rise in drone usage among civilians. For this, drones or swarms of drones are deployed in open ambience, which are prone to various security issues. Blockchain is a budding technology that gives solutions to many such security issues like signal jamming, data integrity, mid-air collision, communication, etc. [22]. Figure 4.3 shows the blockchain-envisioned drone network between healthcare service providers and healthcare service clients.

4.7 CHALLENGES AND OPEN ISSUES FOR FUTURE RESEARCH

4.7.1 Challenges in IoMD for Healthcare Services

Drones are playing a significant role in healthcare-related applications like transporting medical aid in rural areas, saving sinking people, monitoring large gatherings during the pandemic, delivering lab samples for analysis, etc. One must understand that apart from benefits, drones are having challenges comparatively with traditional transport.

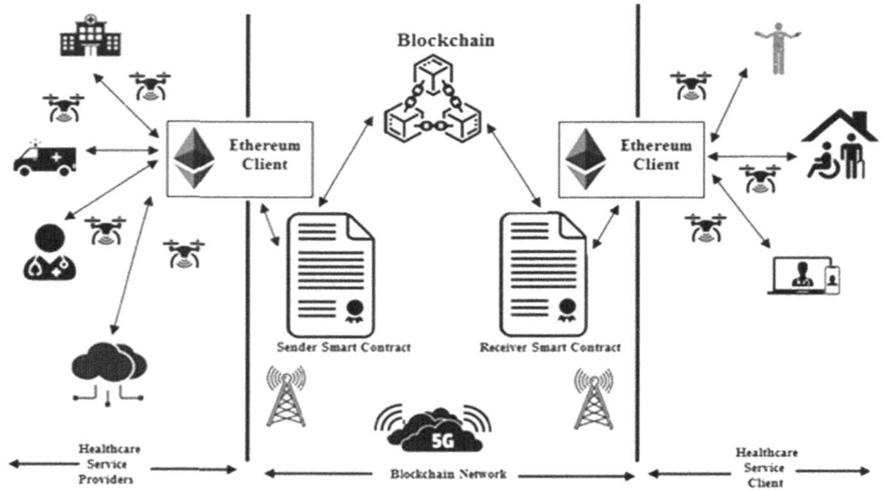

FIGURE 4.3 Blockchain-envisioned drone network.

4.7.1.1 Adoption of IoMD

There are technological, organizational, and environmental challenges possible in drone adoption [23]. The technological challenges are lack of resistance to harsh climatic conditions, limited battery life, limited payload capacity, network delays, limited bandwidth, localization inaccuracy, privacy intrusion, and feelings of fear and insecurity. The organizational challenges are immaturity of drones, lack of financial resources, high investment costs, lack of user skills and expertise, need for training and education, user acceptance, and negative public acceptance. The environmental challenges are lack of regulation, slow and bureaucratic nature of disaster response and regulatory bodies, lack of well-defined standards, accountability and responsibility issues, and regulatory uncertainties.

4.7.1.2 Issues in Security

The adoption of drones may lead to security issues: market availability including terrorists and criminals, weak design that causes security threat, weak policies of authorization and authentication, non-real-time isolation that leads to instant disconnection or turns off the drone, limited testing phase with respect to threat level, inadequate forensics capability, weak level of protection, weak cryptographic suites, and limited frequency range [24]. These security issues can be handled by issuing a drone license for legal registration, firmer restriction to avoid illegal usage of Drones, enhanced surveillance with video feedback, further education to understand threats, firmer law to avoid unauthorized drone usage and for illegitimate operators, defining restricted and confined areas to avoid damage of property and life, enhanced drone detection method to alarm the approach of external drone, lightweight security system for intrusion detection, authentication, and dynamic cryptography.

4.7.1.3 Governance and Regulation on the Internet of Drones Adaption

Most of the drone types need wireless communication to connect with pilots to perform flight. In addition, they need radio communication for cameras and sensors which is possible only by the global frequency spectrum. The usage of frequency spectrum is defined by an international regulation called radio regulation, which contains the list of all frequency allocations [25]. The usage of healthcare drones is allowed in most countries within the visual line of sight (VLS), which is a limited coverage area. In order to provide better healthcare services the law needs to be framed to permit beyond visual line of sight (BVLS) operation. In India, the ministry of civil aviation announced the regulation for drones in December 2018. This regulation insists "No permission No takeoff," which means one-time registration is required for the pilot, the drone, and its owner. The Drone Regulation 2.0 [26] examines:

- The software and hardware of the drone for secure and controlled operation
- Automated operation connected to the framework of air space management
- BVLS operation
- Establishment of global standards
- Modification on civil aviation requirements

The violation of Drone regulation 2.0 may enforce: (a) suspension or cancellation of unique identification number (UIN) or unmanned aircraft operator permit (UAOP) of drone; (b) action as per Aircraft Act 1934; (c) penalties as per IPCs (such as 287, 336, 337, 338, or any relevant section of IPC).

4.7.2 Research Gap and Future Direction on IoMD

Drone technology is a growing technology and needs to mature in the future to face the fear of natural disasters and calamities, and unexpected pandemics like the recent COVID-19. Much research is ongoing to achieve this aim but we still need to understand the current gap and issues with the help of academic investigation. A review on humanitarian drones suggested the future research direction based on drone capability, performance outcome, and technological-organizational-environmental barriers [23]. Table 4.2 lists the academic capabilities, identified gaps, and future research direction.

TABLE 4.2
Research Gap and Future Direction on IoMD

Academic Investigation	Identified gap	Future Research Direction
Capabilities	Transportation and delivery	• Assessing the feasibility of on-time and safety of delivery • Auto identification of the victims • Strategies to find optimal last-mile delivery
	Surveying and monitoring	• Assess the outcome of usage • Effective system to monitor crowded place • Strategy to monitor high-risk place

(*Continued*)

TABLE 4.2 (Continued)
Research Gap and Future Direction on IoMD

Academic Investigation	Identified gap	Future Research Direction
	Communication and integration	• Effective communication in drone and swam of drones • Opportunities to integrate emerging technologies
Performance Outcomes	Flexibility and responsiveness	• Assessing the efficiency • Measuring the performance • Understanding the impact on usage
	Cost reduction	• Assessing cost reduction due to drone integration • Identifying cost optimization model • Minimising routing cost in hybrid transportation
	Sustainability	• Assess social and environmental performance • Empower to provide effective service
TOE Barriers	Technological	• Exploring technological features to increase performance • Self-estimation of remaining flight time, physical resource, path planning • Improve privacy protection • Promoting ethical practice
	Organizational	• Cost benefit analysis on proposing model • Minimize loss in case of crash • Integration of prominent technology
	Environmental	• Propose model to follow rule and regulation on flying drone • Understanding law related to drone implementation

4.8 CONCLUSIONS

Despite the various applications, the evolution of drones in healthcare applications is more challenging due to emergency situations which do not have control of date, time, and place. This study signifies how drones are being explored in healthcare applications. In this chapter, the architecture, applications, tools, and technologies of IoMD; COVID-19 management using drones; recent technologies, issues, and challenges in drone adoption; and future research directions are discussed. However, factors such as governance, legal medical issues, technology adoption, cost, and acceptance are currently challenging the usage of drones.

REFERENCES

[1] W. Fancher, Drones for Medical Supplies, "*Unmanned Drones for Medical Supply Delivery in China Major Qualifying Project*." Apr 2017.

[2] M. Gharibi, R. Boutaba, and S. L. Waslander, "Internet of drones," *IEEE Access*, vol. 4, pp. 1148–1162, 2016, doi: 10.1109/ACCESS.2016.2537208.

[3] M. Angurala, M. Bala, S. S. Bamber, R. Kaur, and P. Singh, "An internet of things assisted drone based approach to reduce rapid spread of COVID-19," *Journal of Safety Science and Resilience*, vol. 1, no. 1, pp. 31–35, Sep. 2020, doi: 10.1016/j.jnlssr.2020.06.011.

[4] S. J. Kim, G. J. Lim, J. Cho, and M. J. Côté, "Drone-aided healthcare services for patients with chronic diseases in rural areas," *Journal of Intelligent and Robotic Systems: Theory and Applications*, vol. 88, no. 1, pp. 163–180, Oct. 2017, doi: 10.1007/s10846-017-0548-z.

[5] G. Ling and N. Draghic, "Aerial drones for blood delivery," *Transfusion*, vol. 59, no. S2, pp. 1608–1611, Apr. 2019, doi: 10.1111/trf.15195.

[6] A. Claesson *et al.*, "Unmanned aerial vehicles (drones) in out-of-hospital-cardiac-arrest," *Scandinavian Journal of Trauma, Resuscitation and Emergency Medicine*, vol. 24, no. 1, Oct. 2016, doi: 10.1186/s13049-016-0313-5.

[7] J. Li, W. W. Goh, and N. Z. Jhanjhi, "A design of iot-based medicine case for the multi-user medication management using drone in elderly centre," *Journal of Engineering Science and Technology*, vol. 16, no. 2, pp. 1145–1166, 2021.

[8] A. Kumar, K. Sharma, H. Singh, S. G. Naugriya, S. S. Gill, and R. Buyya, "A drone-based networked system and methods for combating coronavirus disease (COVID-19) pandemic," *Future Generation Computer Systems*, vol. 115, pp. 1–19, Feb. 2021, doi: 10.1016/j.future.2020.08.046.

[9] M. Kunovjanek and C. Wankmüller, "Containing the COVID-19 pandemic with drones - Feasibility of a drone enabled back-up transport system," *Transport Policy*, vol. 106, pp. 141–152, Jun. 2021, doi: 10.1016/j.tranpol.2021.03.015.

[10] M. Poljak and A. Sterbenc, "Use of drones in clinical microbiology and infectious diseases current status, challenges and barriers," *Clinical Microbiology and Infection*, vol. 26, pp. 425–430, 2020.

[11] Z. Ghelichi, M. Gentili, and P. B. Mirchandani, "Logistics for a fleet of drones for medical item delivery: A case study for Louisville, KY," *Computers and Operations Research*, vol. 135, Nov. 2021, doi: 10.1016/j.cor.2021.105443.

[12] C. Wankmüller, M. Kunovjanek, and S. Mayrgündter, "Drones in emergency response – evidence from cross-border, multi-disciplinary usability tests," *International Journal of Disaster Risk Reduction*, vol. 65, 2021, doi: 10.1016/j.ijdrr.2021.102567.

[13] R. Sokullu, A. Balcı, and E. Demir, "The role of drones in ambient assisted living systems for the elderly," in *Lecture Notes in Computer Science (including subseries Lecture Notes in Artificial Intelligence and Lecture Notes in Bioinformatics)*, vol. 11369 LNCS, Springer Verlag, pp. 295–321, 2019. doi: 10.1007/978-3-030-10752-9_12.

[14] J. C. Rosser, V. Vignesh, B. A. Terwilliger, and B. C. Parker, "Surgical and medical applications of drones: A comprehensive review," *Journal of the Society of Laparoendoscopic Surgeons*, vol. 22, no. 3, 2018. doi: 10.4293/JSLS.2018.00018.

[15] A. Tahir, J. Böling, M. H. Haghbayan, H. T. Toivonen, and J. Plosila, "Swarms of Unmanned Aerial Vehicles—A Survey," *Journal of Industrial Information Integration*, vol. 16, 2019. doi: 10.1016/j.jii.2019.100106.

[16] B. Hiebert, E. Nouvet, V. Jeyabalan, and L. Donelle, "The application of drones in healthcare and health-related services in north america: A scoping review," *Drones*, vol. 4, no. 3, pp. 1–17, 2020. doi: 10.3390/drones4030030.

[17] "COVID-19 Scenario-Emerging Role of Drones in India Recommendations by FICCI Committee on Drones." [Online]. Available: https://www.gpsworld.com/china-fights-coronavirus-with-delivery-drones/
[18] "DRONE TECH and the Introduction of smart drones," Air Drone Craze, 30-Sep-2021. [Online]. Available: https://airdronecraze.com/drone-tech/. [Accessed: 9-Oct-2021].
[19] A. Abdelmaboud, "The internet of drones: Requirements, taxonomy, recent advances, and challenges of research trends," *Sensors*, vol. 21, no. 17, 2021. doi: 10.3390/s21175718.
[20] J. Chu *et al.*, "Machine learning-based dispatch of drone-delivered defibrillators for out-of-hospital cardiac arrest," *Resuscitation*, vol. 162, pp. 120–127, 2021, doi: 10.1016/j.resuscitation.2021.02.028.
[21] C. Guerber, M. Royer, and N. Larrieu, "Machine Learning and Software Defined Network to secure communications in a swarm of drones," *Journal of Information Security and Applications*, vol. 61, 2021, doi: 10.1016/j.jisa.2021.102940.
[22] T. Alladi, V. Chamola, N. Sahu, and M. Guizani, "Applications of blockchain in unmanned aerial vehicles: A review," *Vehicular Communications*, vol. 23, 2020, doi: 10.1016/j.vehcom.2020.100249.
[23] A. Rejeb, K. Rejeb, S. Simske, and H. Treiblmaier, "Humanitarian Drones: A Review and Research Agenda," *Internet of Things (Netherlands)*, vol. 16, 2021. doi: 10.1016/j.iot.2021.100434.
[24] J. P. Yaacoub, H. Noura, O. Salman, and A. Chehab, "Security analysis of drones systems: Attacks, limitations, and recommendations," *Internet of Things (Netherlands)*, vol. 11, 2020. doi: 10.1016/j.iot.2020.100218.
[25] B. Vergouw, H. Nagel, G. Bondt, and B. Custers, "Drone Technology: Types, Payloads, Applications, Frequency Spectrum Issues and Future Developments," 2016, pp. 21–45. doi: 10.1007/978-94-6265-132-6_2.
[26] "Press Information Bureau Government of India Ministry of Civil Aviation Government announces Regulations for Drones." [Online]. Available: https://pib.gov.in/newsite/PrintRelease.aspx?relid=183093

5 CoVacciDrone: An Algorithmic-Drone-Based COVID-19 Vaccine Distribution Strategy

Akash Ilangovan, Sakthi Jaya Sundar Rajasekar, and Varalakshmi Perumal

CONTENTS

5.1 INTRODUCTION

Tackling the COVID-19 pandemic effectively requires taking these two goals into consideration: prevent COVID-19 infections from spreading rapidly via population movement, and immunizing the population at large, enabling security and stability against an infectious environment. Any system that is proposed to manage a pandemic must be compatible for both the goals. Without immunizing the population, there will always be a danger of future infections (Perumal, Narayan & Rajasekar, 2021).

In the limited-resource setting of a rapid onset pandemic, like many countries face today, we cannot divert all our resources to immunization, as the more immediate need is to provide vaccination to areas with high COVID infection rates, to save lives from a fast-moving disease, and to deprive the virus of carriers. Thus, the system we have proposed, CoVacciDrone, tackles both goals with the help of unmanned aerial vehicles (drones), and methods that we have developed to complement the drone-based delivery system (Kumar, Adarsh & Sharma, 2020, A. Waheed and J. Shafi, 2020, Abdelmaboud, 2021).

The main advantage of the drone-based delivery system is to reach people in remote areas, like in mountains, tribal settlements, various rural areas, and

DOI: 10.1201/9781003252085-5

farmland. In rural areas, drones can provide operational and economic benefits, and serve a wide range of medical functions and necessities (Thiels, Cornelius & Aho, Johnathon & Zietlow, 2015, Haidari, Leila & Brown, Shawn & Ferguson, 2016).

The ourworldindata COVID-19 data set shows us that Third-World countries are the most backward in vaccination rates. For example, Egypt has a vaccination rate of 18%, Kenya 6.9%, and Ethiopia 3.1%. We can infer from this that rural and remote areas are in more dire need of efficient vaccination, and this is where the UAV-based approach can make a bigger difference (ourworldindata.org, 2021).

Lack of vaccination lets the disease spread unchecked, as well as hastens the development of new resistant strains (Andre, 2008).

This lets us achieve more thorough immunization, fight spreading infections more vigorously, and adapt to changing requirements and needs in the fight against the pandemic. New variants can be more effectively countered, and vulnerable demographics can more easily be protected (Kumar, Adarsh & Sharma, 2020, Syed Mohammed, thehindu.com, 2021).

Drone design and payload must also be appropriately concluded for any UAV-based operational system. In our chapter, we will consider the UAVs to be quad-copter drones, with high payload capacities, and with max speeds of 60 to 75 km/h. The drones will be connected to a satellite that can provide GPS functionality, as well as be equipped with LIDAR sensors, allowing it to navigate unusual terrains and detect obstacles (Dorling, Kevin & Heinrichs, 2016, Sorbelli, Coro, Das, 2021).

We imagine the landscape of the pandemic as an enormous amount of graphical data. We need to be able to manipulate this graphical data to suit our needs, in quick timespans, and with reduced resource expenditure. Thus, we introduce the concept of dynamic connectivity, an understanding of the data as an ever-changing, flexible graphical representation, which helps us divert our resources better (Riehl, James & Collins, 2009, Berg, Vheong, Kreveld, 2008).

The proposed system will enable an easy shifting of strategies: from conducting a thorough immunization to countering a fierce outbreak by focusing our resources and preventing disasters like COVID waves that may arise from unexpected mass infections.

Our landscape is first divided into sections. Initially, sections can be used to represent state- or province-level jurisdictions that are commonly found in many countries, that have a federal form of government. This lets the initial monitoring, analysis, resource acquiring, and immunizing be organized by state governments and other similar organizations. As the pandemic intensifies, however, the graphical representation shifts, and the jurisdictions are blurred.

The system enables high-level coordination between different organizations, based on urgent requirements, and the general need to immunize. Each section is one connected component of the graphical representation. The next smaller partition in the hierarchy will be called a COVID division. This is based on the location of Vaccine Distribution and Storage Centers (VDSC). These VDSCs are used to store vaccines in bulk, and to organize drone movement throughout each of their corresponding COVID division. Vaccines are transported from the VDSCs, in either trucks or drones and used to immunize the COVID division. As the division itself is being partitioned, we can prioritize hotspots, down to housing blocks, thus letting us prevent rapid outbreaks effectively. Sections may also accept vaccines from each other, when it becomes more efficient to do so.

During a furious pandemic, resources can be limited in affected areas, and areas that possess surplus can provide lifesaving assistance. Thus, it is not effective to maintain the sections as they are initially. Our proposed systems let us switch to different sections or ask for assistance from different sections. The final level of classification is from COVID divisions to COVID localities. Here, too, Voronoi diagrams are used to sort according to Locality Retrieval Centers (LRCs) and then dynamic graphical structures are implemented (Yang, Liang & Qi, 2015, Anderson, Vegvari, 2020).

The dynamic graphical structures can be represented via two different methods: the Euler tour trees method or the cut-set method. Both methods can provide fast manipulation of graphical data. The description and advantages of these methods will be discussed in the upcoming sections.

This chapter proposes and explains different algorithms and methods for efficient distribution and delivery of vaccines with drones. In 6.3, we explore Voronoi diagrams that can divide a landscape into neighborhoods, which are determined by vaccine distribution centers. In 6.4 to 6.6, we explore dynamic connectivity, a concept that is used to manage ever-changing graphical data in large volumes. Two different methods of dynamic connectivity, Euler tour trees and cut-sets, are explained. Finally, in 6.7, a comparison and analysis between the characteristics of the two methods are given.

Our objective is to find a balance between the goals of total immunization and to counter rapid outbreaks that can change the fate of the pandemic. The variability of the pandemic means that the dangers might be different in nature in various parts of the country. We must both fight these dangers and not let them interfere with our immunization drive, which ensures that the pandemic is done, once and for all. Our proposed system will help us do that (Haidari, Leila & Brown, Shawn & Ferguson, 2016, Sarkodie, Owuse, 2020, Shereen, Khan, Kazmi, 2020).

5.2 DIVIDING SECTIONS INTO COVID DIVISIONS WITH VORONOI DIAGRAMS

A particular federal jurisdiction or section can be divided into COVID divisions with the help of Voronoi diagrams. A COVID division is an area associated with any one of the various Vaccine Distribution and Storage Centers (VDSC) within a particular section. A COVID division usually consists of several localities. When dividing a section into COVID divisions, the following criteria is followed: given a Locality Li, which VDSC is it the closest to? To sort Localities into COVID divisions, Voronoi diagrams are computed (Berg, Vheong, Kreveld, 2008).

Given a set of points, P = {p1, p2, ... Pn}, lying in a region R, every point in R is sorted into a division Di, where every point in Di is closer to pi than all other points in the set P. The following diagram looks like Figure 5.1, as shown.

In Figure 5.1, the landscape has been divided into nine alphabetically ordered regions. In each region Rx, every point in Rx is closer to the point X that denotes the region (e.g., Ra is associated with A, Rb is associated with B, and so on), than

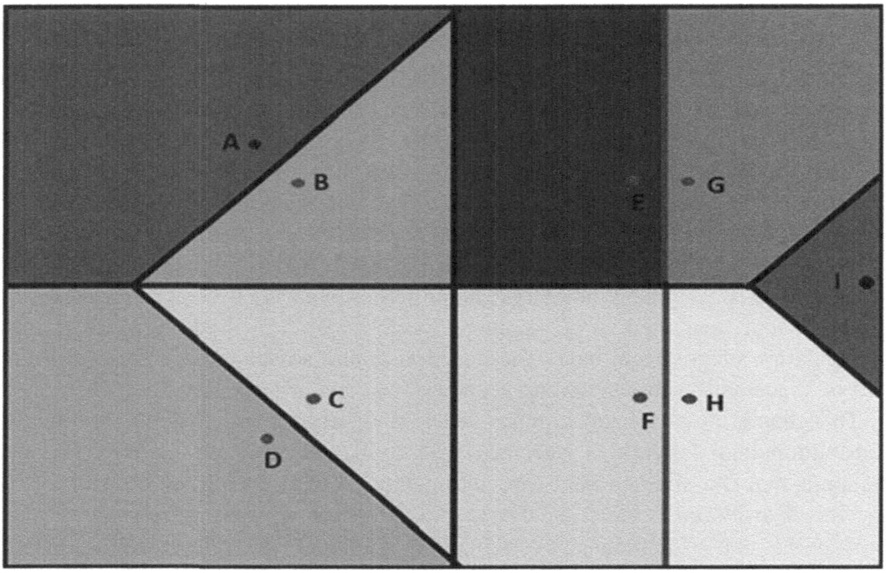

FIGURE 5.1 Voronoi diagram.

any of the other eight points in the region. The advantage of a Voronoi diagram is that each point in the region is associated with the VDSC located closest to it (Tong, Han & Wen Chao, 2012).

When using drone delivery, the cost of traveling from any VDSC to a point p is equal to the Euclidean distance of p from that VDSC. Thus, the usage of Voronoi diagrams minimizes the cost of travels taken by drones as it is a resource constraint device. A Voronoi diagram is constructed by the following algorithm:

CONSTRUCT VORONOI: ALGORITHM

1. **For** each point pi,
2. Extend the perpendicular bisectors for pi and each of its neighboring points.
3. The region formed by intersection of perpendicular bisectors around it is the corresponding Voronoi region.

A Voronoi diagram can be constructed in O (n log n) time.

To divide a section into COVID divisions and to divide COVID divisions into COVID localities, the CONSTRUCT VORONOI algorithm is used, to minimize the cost of drone usage.

5.3 COST OPTIMIZATION WITH DYNAMIC CONNECTIVITY

There are several factors that must be considered during CoVacciDrone usage during a pandemic. These factors are:

1. The traveling capacity of drones.
2. The expiry period of vaccines.
3. The amount of vaccine stock
4. Cost of transporting the vaccines.

The final factor, the cost of acquiring vaccines relies upon a set of conditions that change throughout the course of the pandemic. When the stock of vaccines is limited, it is better to acquire surplus vaccines from nearby COVID divisions, or nearby COVID localities, instead of sending orders to VDSC, which can strain the supply to the VDSC and also be more time consuming. When we simply acquire vaccines from nearby COVID divisions, or nearby COVID localities, the distance of vaccine transport can be reduced by hundreds of kilometers. When we face scenarios like this, it is computationally expensive to calibrate new routes and find viable sources. Thus, it is useful to maintain a statistical representation of vaccine availability, which can be modified easily. We can accomplish this by maintaining **dynamic graphs.** Dynamic graphs are graphical representations of data that can be modified with computational ease (Wikipedia, Dynamic connectivity).

They can be used to quickly compute viable sources for acquiring vaccines. There are diverse ways of representing dynamic graphs, each with different advantages. Dynamic graphs are maintained in CoVacciDrone in two levels: first considering the VDSCs for each COVID division as nodes and then by considering the Locality Receiving Centers, or LRC for each COVID locality as nodes on the lower hierarchical level. Dynamic graphs are used in both levels, in conjunction with Voronoi diagrams, to create a statistical representation of vaccine availability, and requirement, which can be optimized as conditions change.

5.4 DYNAMIC CONNECTIVITY WITH EULER TOUR TREES

Dynamic connectivity can be achieved by using the following data structure: Euler tour tree or ETT. A Euler tour tree can be constructed using the Kruskal method for constructing minimum spanning trees (stanford.edu, Lacki, 2015).

Each connected component in a graph is represented by a single ETT. Thus, the whole graph, which represents all sections, with each section as a connected component, and with COVID divisions as nodes, can be represented by a forest of ETTs. A possible section can be represented as a connected component like in Figure 5.2.

Figure 5.2 shows how the highest level of division, a section, can be represented as a connected component. As seen in the previous section, dynamic connectivity is essential to quickly find viable vaccine sources and optimize delivery time. In our

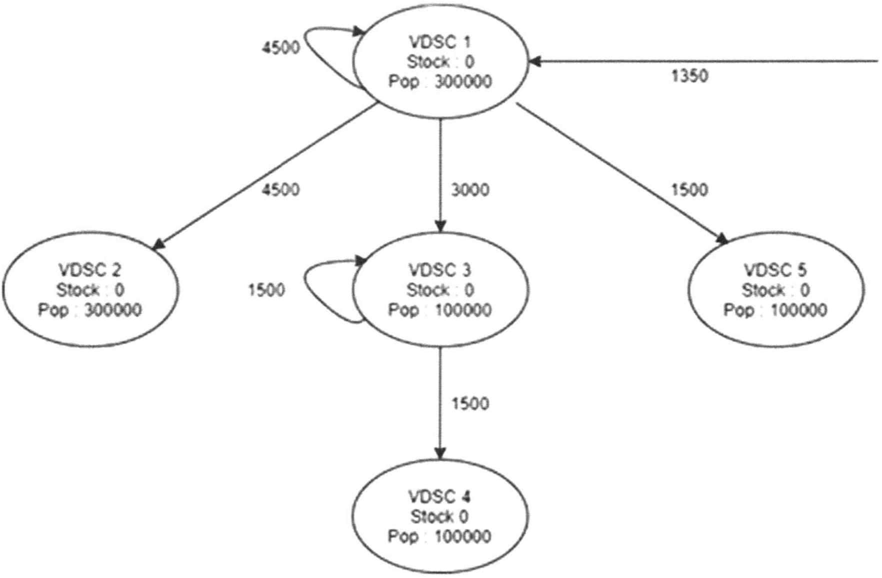

FIGURE 5.2 Section with 5 VDSCs.

model, dynamic connectivity can even be maintained within a COVID division, between localities. Thus, our model has two layers of ETT forests. First, we shall see how to construct an ETT forest. To do this, we make use of Kruskal's method for finding a minimum spanning tree.

KRUSKAL MST:

1. Sort all edges in non-decreasing order.
2. Pick the smallest edge. If it does not form a cycle with the edges chosen so far, keep the edge.
3. Do until n-1 edges are present (n is total no of vertices)

(Horowitz, Sahni, Rajasekaran, 2007, Khaledi, Afghah, Razi, 2018)

The output of Kruskal's method is a forest of trees, each tree representing a single connected component. Now, we shall see how dynamic connectivity is achieved with ETTs.

Initially, we create a forest of ETTs by considering all COVID divisions as nodes, and each section (federal jurisdiction)as a connected component. Remember that each state is assigned a number of VDSCs. We construct Voronoi diagrams under these VDSCs, to sort a section into COVID divisions. Now how do we create a vaccine distribution algorithm? To do this, let us consider the factors of requirement and availability. When the requirement is low, and availability is high, we have a surplus. Let us assume that a VDSC, V1, has a surplus. Then V1 can simply distribute its vaccines to its neighboring COVID divisions, by dividing its vaccine stock in the ratio of populations of its neighboring COVID divisions. Any neighboring VDSC, say V2, can distribute the vaccines it receives from V1 to its neighbors. It can also take its own populations into account, if the COVID division corresponding to V2 has COVID-positive patients and needs vaccines for its own COVID division. Now, what if the neighbors of V1, and V1 itself, have no cases? Then the vaccines can be distributed by population to all of V1's neighbors, their neighbors and so on. In this way, COVID eventualities can be prepared for, while at the same time, infection spread can be resisted, by providing urgently required vaccines, to hotspot areas. If the use of the minimum spanning tree is to balance the availability and requirement factors, then where does dynamic connectivity benefit us? We have discussed this in the previous section. If V1 is in urgent need of vaccines, and cannot acquire them from its own section, then it must request vaccines from a VDSC in a different section. Also, if available vaccines are closer in a VDSC in a different section, then to minimize cost, vaccines can be acquired from that VDSC. Thus, edges must be added to the forest, and connected components must be merged.

Similarly, if a COVID division, has been completely vaccinated, it is better to remove that node from the forest, as it is no longer required as an intermediate. However, when a drone has reached its flight capacity, it must find nearby intermediate points, and at that point, previously computed forests can be useful. Thus, a section should store the parts of the forest which correspond to its connected component, as they may prove useful later.

Now we must create an algorithm that manages the ETT forest and adapts to changing conditions.

MAINTAIN ETT:

1. **For** a VDSC, Vi,
2. **If** Vi gets a request from outside the section, from VDSC Vj, add edge (Vi, Vj).
3. **If** Vi is fully vaccinated, delete all edges from Vi.
4. **If** Vi has a surplus of vaccines and a fully vaccinated population, distribute to its neighbors in the section by ratio of their respective populations.
5. **Else If** Vi has a surplus of vaccines and no active COVID cases, but its neighbors do, and do not have surplus, distributed by ratio of population.
6. **Else If** Vi has a surplus of vaccines, and no active COVID cases, and its neighbors also do not have active cases, and no surplus, distributed by ratio of population, including its own population.
7. **Else If** Vi has a stock of vaccines, and active COVID cases, and its neighbors also have active cases, and no surplus, distributed by ratio of population, including its own population.
8. **Else If** Vi has a stock of vaccines, and active COVID cases, and its neighbors have no active cases, and surplus, distribute all vaccines within itself.

9. **Else If Vi** has no stock of vaccines, and active COVID cases, request vaccines, from outside and inside the section.
10. **If** the closest VDSC, Vk, that responds to Vi's request is in a different section, then add (Vi, Vk) and delete all edges of Vi to its former section.
11. **Else** when Vi has no stock and no active cases, request vaccines from inside the section.

Now that we have written an algorithm for dynamic connectivity, we must write algorithms for adding and deleting edges. First, we must understand the concept of specificity. Initially all edges have the specificity of 0. If an edge is deleted, then the ETT is split in two. The smaller ETT has all its edges and their specificities increased by 1. We maintain a total of Log n forests, where each forest Fi contains all the edges having specificity greater than or equal to i.

0 <= i <= log n – 1. We can see how specificities are changed in Figure 5.3.

The following algorithms are used to add and delete edges,

ADD (Vi, Vj)

1. Add Vi,Vj in F0 only.

DELETE (Vi, Vj)

1. **For** x varies from 0 to specificity (Vi, Vj) = k
2.   Remove (Vi, Vj) from Fi
3. Ti = Larger Split Tree, Tj = Smaller split tree
4. **For** i = k to 0
5.   **For** each edge in Fi that is present in Tj
6.   **If** the edge is not incident to Ti
7.   Increase specificity by 1.
8.   Push edge to Fi+1
9.   **Else**
10.   Add the edge from F0 to Fi.
11.   Exit.

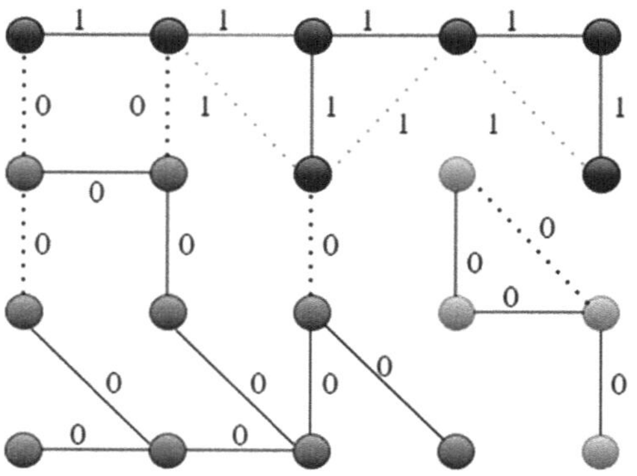

FIGURE 5.3 Changing specificities in an ETT forest.

5.5 DYNAMIC CONNECTIVITY WITH CUT-SETS

There are some improvements that can be made to the Euler tour tree, or ETT, method. A method that can improve the computational expenses, required by the ETT method, is the cut-set method (Lacki, 2015, Wikipedia, Dynamic connectivity).

Given a graph G (V, E), cut-set(T) is defined as the set of edges that connect T with V/T, where T is a subset of V. A cut-set is represented in computation by two fields: first, a set of all edges included, and then the XOR of all the edges present in the cut-set. The cut-set operations can be used to provide dynamic connectivity, i.e, insert and delete edges in G.

First, let us try to visualize a cut-set. Each vertex in G is represented by a binary number. Then we know that each vertex can be represented by Log n bits, where n is the number of vertices. Next, each edge is represented by appending the bits that represent its two (2) vertices together; so, an edge is represented by $2 * \text{Log } n$ bits. The second component in representing a cut-set is it's XOR, where XOR(T) is the XOR of all vertices in T. First, for a vertex *v*, we compute XOR(v) = XOR of all the edges adjacent to v. XOR of all these values gives us XOR(T).

The number that is used to represent an edge e, occurs twice in XOR(T), if both end points of *e* lie in *T*. Thus, we reach the valuable conclusion: If *e* has both its endpoints in *T*, it vanishes and has no effect on XOR(T). Based on this fact, we can assert the following:

1. If XOR(T) = 0, then there is no edge in T that connects T to the rest of V. That is, T cannot be united with V.
2. If XOR(T) = e, where e is any edge, then e is the only edge that combines T with V.
3. The third case is when XOR(T) is a non-zero number that cannot be identified with any single edge. This means there are more than one outgoing edges from T, but they cannot be identified properly.

For this third case, we can perform the following: first create Log n levels of the cut-set structures. For each cut-set, the first level, L0 has all the edges present, the next-lower level L1 has an arbitrarily chosen half of the edges in the just upper-level L0, L2 has half the edges chosen arbitrarily from L1, and so on. If we were to stop with this level of preprocessing, then we would do the following: XOR the level L0, and see if a single edge is outputted, if not try L1, then L2, and so on. The probability of success is only 1/9. So, we make further improvements: create horizontal levels, in addition to the vertical. Create C * Log n versions of the vertical level structure horizontally. Each level has an arbitrary reduction in edges from its nearest left level. Now we can query the XORs of all the levels. The probability that we fail can be made to 0, if we chose C appropriately.

We can now maintain dynamic connectivity by keeping a single forest (no specificities required) and the various levels of the cut-sets. We use Kruskal's algorithm to create a spanning forest. Then, for each of the trees in the forest, create the corresponding cut-set field. We can perform the insert and delete edges operations by the following algorithms. Keep F as the forest.

ADD (Vi, Vj)

1. //First add the edge to the forest and modify the forest accordingly.
2. **For** each edge e incident on Vi from Vk
3. **if** cost(Vi, Vk) > cost(Vk, Vj) + cost(Vj, Vi)
4. Add (Vk, Vj) , (Vj, Vi) to F
5. **For** each edge e incident on Vj from Vn
6. **if** cost(Vj, Vn) > cost(Vn, Vi) + cost(Vi, Vj)
7. Add (Vn, Vi) , (Vi, Vj) to F
8. //Next XOR the edge with the cut-set(s)
9. Add (Vi Vj) to the required cut-sets.// All cut-sets that contain Vi and Vj
10. If two components are united, merge the required cut-sets.

DELETE (Vi, Vj)

1. Delete (Vi, Vj) from the spanning forest.
2. Delete (Vi, Vj) from all cut-sets that contain either Vi or Vj
3. Do XOR of the rectangular set of layers for T1.
4. Merge cut-set(T1) with Cut-set(Ti).//Ti is the incident component for any edge obtained from the XOR
5. **For** all Ti
6. ADD (Vk, Vj).//Where Vk from T1 and Vj from Ti

In this way, adding and deleting edges can be performed with cut-sets. Cut-sets can be used as one of the layers for the CoVacciDrone system, where either the smaller division: from COVID divisions to COVID localities: or the larger division: from sections to COVID divisions can be performed. In this way, the statistical stability of the ETT can be combined with the speed of the cut-set method.

5.6 RESULTS AND OBSERVATION

After constructing different variations of our model, we can provide conclusions about the advantage of dynamic graph models, the comparisons between them, and any shortcomings faced by them. It is observed that dynamic graph models provide for an elevated level of flexibility that helps organizations facilitate care and resource allocation to vulnerable communities. Given the massive size of the graph data set for a country like ours, it is crucial that manipulations are allowed in short spans of time. Indeed, our algorithms provide reasonable computational complexity, around $O(poly(log\ log\ n))$, that is manageable, when n can be in the order of hundreds of thousands , when you consider the localities in our country. The algorithms proposed also allow organizations like state governments to supervise operations at the initial stage, and foster cooperation at later stages. Table 5.1 gives an overview of the computational complexities of the two main dynamic algorithms.

We can see from Table 5.1 that the cut-set method is faster, with a worst-case time complexity of $O(poly(log\ log\ n))$. In the ETT method, the amortized time complexity is itself only $O((log\ log\ n)^2)$. Given the large volume of data, space complexity is also a valuable consideration. The ETT method is shorter by two

TABLE 5.1
Overview of Time and Space Complexities for two Dynamic Graph Algorithms

	Storage Complexity	Amortized Time Complexity for Operations	Worst Case Time Complexity for Operations	Construction Time Complexity
ETT Method	$n \times log\ log\ n$	$(log\ log\ n)^2$	n	$n \times log\ log\ n$
Cut-set Method	$(n \times n)^3$	n	$poly(log\ log\ n)$	$n \times log\ log\ n$

whole orders of Log n. This tells us that for the first division, which is from section to COVID divisions, the number of operations is much less than the second COVID locality division. Thus, a structure that makes use of both algorithms could use the ETT method for the top layer and the cut-set method for the lower layer. The methods we have proposed in this chapter function as a solid foundation for many more functional components. IoT can be added to coordinate with drones, or to collect medical data. Sanitization can be done electronically, to shield against infection in hotspot areas. Big data can be used to collect vaccine reaction data and protect target groups (Rosser, James & Vudatha, 2018, Rajasekar, 2021).

The CoVacciDrone system highlights the importance of using high-level computational techniques against pandemics, a strategy that could prove to be a game changer in developing countries.

REFERENCES

Abdelmaboud, A. The Internet of Drones: Requirements, taxonomy, recent advances, and challenges of research trends. *Sensors*. **21**(17): 5718 (2021).

Adnan Shereen, M., Khan, S., Kazmi, A., Bashir, N., and Siddique, R. COVID-19 infection: origin, transmission, and characteristics of human coronaviruses. *Journal of Advanced Research*, 24: 91–98 (2020). doi: 10.1016/j.jare.2020.03.005.

Anderson, R. M., Vegvari, C., Truscott, J., and Collyer, B. S. Challenges in creating herd immunity to SARS-CoV-2 infection by mass vaccination. *Lancet* **396**: 1614–1616 (2020).

Andre, F. E. *et al.* Vaccination Greatly Reduces Disease, Disability, Death and Inequity Worldwide. Bulletin World Health Organization **86**: 140–146 (2008).

Berg, M. D., Vheong, O., Kreveld, M. V., and Overmars, M. Computational Geometry Algorithms and Applications, 3rd Edition, Springer, (2008).

Dorling, K., Heinrichs, J., Messier, G., and Magierowski, S. Vehicle Routing Problems for Drone Delivery. IEEE Transactions on Systems, Man, and Cybernetics: Systems. **47**: 1–16 (2016). 10.1109/TSMC.2016.2582745.

Haidari, L., Brown, S., Ferguson, M., Bancroft, E., Spiker, M., Wilcox, A., Ambikapathi, R., Sampath, V., Connor, D., and Lee, B. The Economic and Operational Value of Using Drones to Transport Vaccines. *Vaccine* **34**: 4062–4067 (2016). https://doi.org/10.1016/j.vaccine.2016.06.022.

Horowitz, E., Sahni, S., and Rajasekaran, S. *Fundamentals of Computer Algorithms*, 2nd Edition, University Press, (2007).

Khaledi, M., Rovira-Sugranes, A., Afghah, F., and Razi, A. On Greedy Routing in Dynamic UAV Networks. 2018 IEEE International Conference on Sensing, Communication and Networking (SECON Workshops), (2018), pp. 1–5, doi: 10.1109/SECONW.2018.8396354.

Kumar, A., Sharma, K., Singh, H., Naugriya, S., Gill, S. S., and Buyya, R. A Drone-based Networked System and Methods for Combating Coronavirus Disease (COVID-19) Pandemic. Future generations computer systems: FGCS **115** (2020). 10.1016/j.future. 2020.08.046.

Lacki, J. Dynamic Graph Algorithms for Connectivity Problems. Institute of Informatics, University of Warsaw, (2015).

Perumal, V., Narayanan, V., and Rajasekar, S. J. S. Detection of COVID-19 using CXR and CT images using Transfer Learning and Haralick features. Applied Intelligence. **51**: 341–358 (2021).

Rajasekar, S. J. S. An Enhanced IoT Based Tracing and Tracking Model for COVID-19 Cases. SN Computer Science **2**: 42 (2021).

Riehl, J., Collins, G., and Hespanha, J. Cooperative Search by UAV Teams: A Model Predictive Approach Using Dynamic Graphs (2009).

Rosser, J., Vudatha, V., Terwilliger, B., and Parker, B. Surgical and Medical Applications of Drones: A Comprehensive Review. JSLS: Journal of the Society of Laparoendoscopic Surgeons **22**: e2018.00018 (2018). 10.4293/JSLS.2018.00018.

Sarkodie, S. A. and Owusu, P. A. Global Assessment of Environment, Health and Economic Impact of the Novel Coronavirus (COVID-19). Environment, Development and Sustainability (2020). doi: 10.1007/s10668-020-00801-2.

Sorbelli, F. B., Corò, F., Das, S. K., and Pinotti, C. M. Energy-Constrained Delivery of Goods With Drones Under Varying Wind Conditions. *IEEE Transactions on Intelligent Transportation Systems* **22**(9): 6048–6060 (Sept. 2021). doi: 10.1109/TITS.2020. 3044420.

Thiels, C., Aho, J., Zietlow, S., and Jenkins, D. Use of Unmanned Aerial Vehicles for Medical Product Transport. Air Medical Journ*al* **34** (2015). 10.1016/j.amj.2014.10.011.

Tong, H., Wen Chao, W., Qiang, H., and Bo, X. Path planning of UAV-based on Voronoi diagram and DPSO. *Procedia Engineering* **29**: 4198–4203 (2012). 10.1016/j.proeng. 2012.01.643.

Waheed, A. and Shafi, J. Successful Role of Smart Technology to Combat COVID-19. 2020 Fourth International Conference on I-SMAC (IoT in Social, Mobile, Analytics and Cloud) (I-SMAC), (2020), pp. 772–777, doi: 10.1109/I-SMAC49090.2020.9243444.

Yang, L., Qi, J., Xiao, J., and Yong, X. A literature review of UAV 3D path planning. *Proceedings of the World Congress on Intelligent Control and Automation (WCICA)* **2015**: 2376–2381 (2015). 10.1109/WCICA.2014.7053093.

https://en.wikipedia.org/wiki/Dynamic_connectivity

https://ourworldindata.org/covid-vaccinations

https://web.stanford.edu/class/archive/cs/cs166/cs166.1146/lectures/17/Small17.pdf

https://www.thehindu.com/news/cities/Hyderabad/uav-that-can-deliver-heavier-payloads-to-longer-distances/article32518295.ece

6 Ambulance Drone for Rescue – A Perspective on Conceptual Design, Life Detection Systems, and Prototype Development

S. Mano and V.M. Sreehari

CONTENTS

6.1 INTRODUCTION

An unmanned ambulance drone (UAD) is a custom-designed drone for casualty evacuation. The ambulance drone is a specially designed unmanned aerial vehicle (UAV) that can perform rescue operations. Consider a situation like an army man who was injured seriously in a battlefield and he needs to be rescued from there immediately as the doctors are unable to reach the location. In such cases, an ambulance drone discussed here will be able to lift the injured army

DOI: 10.1201/9781003252085-6

man and transport him back to base. The other feasible practical applications are:

- Rescue operations during a fire hazard in a high-rise building.
- Civilian, cargo, and organ transportation.
- Soldier troops and cargo transportation.
- Monitoring wind turbine maintenance.
- Emergency situation in offshore oil rigs and ships.
- Agriculture sprayer.
- Forest fire extinguisher.

The ambulance drone, done by TU Delft University [1], is a compact flying toolbox containing essential supplies for advanced life support. A health-integrated rescue operations [2] drone system delivers a case that includes medical supplies. A person near the stricken patient is expected to put on the glasses, which sends the video in front of them to a remote physician. The physician can then see what's going on and lead the deputized civilian through the necessary treatment steps that utilize the supplies in the case. Another medicine delivery drone performed an extensive flight test in Davis, California, Zipline [3], packages the order, and then launches it into flight, racing along over 100 km/h. Due to technological progresses in batteries, electronics, and sensors, high-performance multi-rotors can now be manufactured at a moderately low cost [4]. But rather than delivering medical supplements, the rescue itself needs to be done, employing UADs in several cases, as mentioned earlier.

The UAD may be routinely flown to any emergency scenario and utilized to lift, carry, and medicate anybody in want of assistance. It's a drone ambulance that could perform in all conditions. The UAD is a vertical take-off and landing (VTOL) aircraft; this means that it can be lifted and landed in open vicinity close to the patient. In the occasion of an emergency, the proposed drone is specifically tailor-made to deliver. It consists of a compact virtual controller, a compact defibrillator, medication, a stretcher, an oxygen ventilator, and cardiopulmonary resuscitation equipment, in addition to different crucial substances to apply, even while being transported during an emergency. The UAD is a specifically designed UAV that transfers injured or deceased human beings in scientific emergencies over distances or terrain that might not be possible for a conventional ground ambulance. Because of their emergency nature, such UADs will be given precedence in sky routes [5]. In certain circumstances, a UAD capable of traveling, lifting an injured soldier (unconscious), and transporting them back to base is necessary.

In recent years, the drones or UAVs support a medical field during emergency in several ways, like providing an instant network infrastructure and medical first aid service and it is employed to deliver medicine to patient, etc. The Drone Civil Air Patrol Wing organization is developed for rapid processing of permission for emergency situations like promotion of industry expansion and public awareness. Importantly, public participation must be emphasized for these to become routine processes [6–8].

The authors have initially studied the conceptual design [9]. This present chapter discusses the conceptual design; discussion on life detection systems; and prototype production of an UAD, which is an ambulance that can be flown (low altitude medium range) to any emergency and lift, carry, and medicate patients. The various systems needed for UADs are addressed initially and the prototype development from the conceptual design stage is discussed finally. The perspective of systems like life detection system, robotic arm, etc., are incorporated into the UAD discussed in the present chapter, which makes the UAD significant and highly relevant. The UAD has human presence detection systems and robotic arm technologies, both of which are critical in a UAD. The proposed conceptual ambulance drone houses systems for detecting human presence (impulse-based Doppler radar, thermographic camera, ultra-wide-band (UWB)). Anyhow, these life detection systems have not yet been incorporated into the present working model. Also, after discussing the systems, a perspective on prototype development of such UADs is discussed. The ambulance drone is also incorporated with two axial gimbals systems to the fuselage (where the patient is being carried to the hospital) in order to provide the opposite forces/moment to the force/moment created by the drone while cruising to balance it. Thus, the patients inside the fuselage do not have the airsickness problem. This tilt mechanism provides more stability and prevents drones from tilting forward or backward during the cruise. Wheels are used in places where a rotor has a problem operating due to space constraints, as in dense urban areas with tall buildings. The ambulance drone is incorporated with robotic arm technology to lift humans. Robotic arm technology is incorporated for causality evacuation and is made of glass-epoxy composites and the fuselage structure was made of 3D printing technology in the prototype. Thus, prototypes of UADs were fabricated and tested.

6.2 CONCEPTUAL DESIGN: MAIN COMPONENTS

The low-altitude UAD that would save lives in areas like warfare zones, roads, waterways, etc. Its lifting capability, including payload, is 1,000 kg. Figure 6.1 indicates the proposed conceptual design of interior of a UAD.

The main components are:

A. Fuselage: The cockpit and the stretcher base are in the fuselage. The fuselage houses casualty and life-saving equipment for quick medical assistance. The stretcher platform in the fuselage can be changed with seats, and the plane can be used to transport soldiers in battle, their weapons, and food, as well as people in towns when needed.
B. Main frame: The mainframe supports the fuselage and is made up of hollow cylindrical rods. It has motors on each end. The UAD is lifted and flown by 16 motors that give thrust.
C. Cock pit and controls: Autopilot mode can be employed to fly the UAD.
D. Engine and propeller system: The UAD has 16 electrical engines mounted coaxially at the end points of the axis.

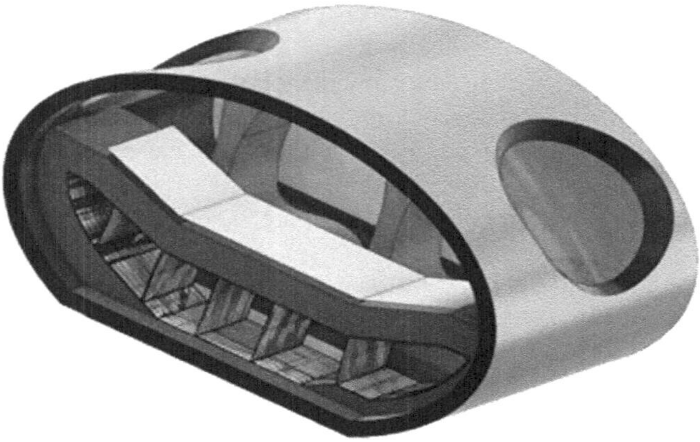

FIGURE 6.1 Conceptual design of interior of UAD.

Preliminary calculations are done and the structural calculations (for the dimensions as in Table 6.1) are performed initially, as in Table 6.1, by considering an elliptical structure [10]. Weight calculations for the proposed UAD are done and significant values are: structural weight = 650 kg, payload weight = 300 kg, miscellaneous weight = 50 kg, So, total weight = 1,000 kg. Propulsion systems, aerodynamic, and performance calculations are performed, and significant values are given in Table 6.2.

TABLE 6.1
UAD Fuselage Specifications

Parameter	Specifications
Fuselage major axis	2,438 mm
Fuselage minor axis	1,524 mm
Fuselage height	1,829 mm

TABLE 6.2
Important Propulsion, Aerodynamic, and Performance Characteristics

Propulsion	Aerodynamics	Performance
Power required for level flight = 105.67 kW	Cruise velocity = 15 m/s	$C_L/C_D = 0.2$
$C_t = 0.0071$	$C_L = 0.371$	Range = 35 km
Thrust = 704.5 N	$C_D = 1.86$	Endurance = 166.6 Min

6.3 SYSTEMS FOR SEARCHING HUMAN PRESENCE – A SURVEY

A rescue operation is one of the regions that attracts a major interest in innovative research. According to the literature survey, the human identifying strategies have been indexed as follows.

6.3.1 Human Presence Detection Employing MMW

This UAD design is proposed to include a revolutionary enhanced approach for human proximity discovery using disguised millimeter-wave (MMW) sensors. In the MMW radiometry sensor, the approach distinguishes between stationary and moving object background techniques. Ka-band radiometers are used within the general energy mode in addition to the connection mode, which responds greatly to self-luminous substances. Figure 6.2 depicts the radiative power emitted through a human, in addition to the general energy. Two self-sufficient splendid hetero-dyne recipients with reception apparatus segregated through a gauge make up the system's back and front equipment. The RF focuses on recurrence, oscillator of neighborhood, and recurrence of IF [11]. The MMW sensors are recommended to be equipped with this proposed UAD and it always checks the prevailing environment.

6.3.2 Method of Identifying Humans from Other Moving Creatures

In the defense sector, it is still a big challenge of identification of object movement, rescue operation, and finding enemy activities. Thorough advanced image processing techniques were used in multi-dimensional image processing technology, method of

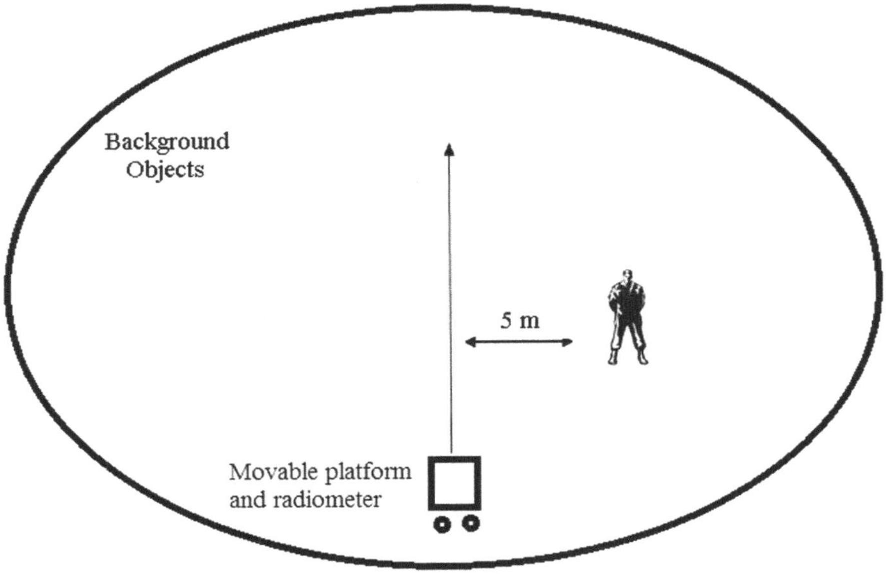

FIGURE 6.2 The MMW sensor setup and its perspective.

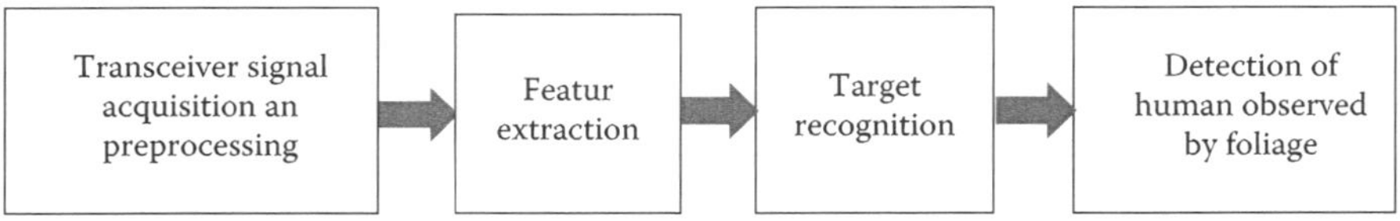

FIGURE 6.3 Identification of human location using UWB.

pattern recognition, and rapid information handling [12]. This specific UAD design is completely applicable for outdoor surveillance and intruder detection. The proposed UAD is designed with UWB radar technology, which is applicable to detect human or other objects by using the echoes from brief span beat transmission. The UWB radar configured with constant wave radars. The UWB radar can be used to classify people based on their breathing, cardiovascular aspects, etc. Because puppies, for example, have the same respiratory movement as humans, the UWB radar should be able to differentiate not just humans but also animals. Subsequently, some other techniques are also proposed in UAD for recognizing those living objects.

In recent years, numerous scientists approached many advanced methodologies for detecting human location (Figure 6.3) and they identified two different methods [13]. The first method uses advanced sensors to identify people, which include both picture and non-picture sensors. Although the video sensors provided a good outcome in terms of arrangement precision, this strategy consumes more power and necessitates additional memory storage devices. Similarly, non-picture sensors with improved techniques such as acoustics mode, seismic ultrasonic, attractive, and advanced infrared sensors are commonly used because of their low power operation and are more cost effective than a picture sensor. This method has an accuracy of 100% for non-target, 96.5% for humans, 96% for other creatures [13].

6.3.3 The Image Processing by Using Thermal Image Technique – Haar-Cascade Classifier

An identity of human looks may be characterized because of the obstacle of all gadgets in a given picture/image. The Haar-Cascade approach is the one that had been a normally implemented approach to pick out the gadgets [14]. These identity strategies consist of four steps to pick out the object. i.e., essential picture, Haar like highlights, flexible boosting, and direction classifier mix. A crucial picture is a way of quickly estimating the Haar-like component by turning one pixel into another. Every pixel in the required image has a value that is derived from the pixel position of the picture. Thermal image processing can be done. Finally, the image or photograph will be created using the heat energy emitted by the item.

6.3.4 The Random Human Body Motion

If wounded soldiers may be recognized quickly, their rescue and the performance with which they are overseen might increase. The wounded soldier body can be found by a proposed UAD by using thermal image cameras. It will determine if the person is alive or dead based on the respiratory system [15–17]. CW radar analyzes the human frame's chest growth due to breath to evaluate the breath rate. The radar cluster can detect turbulence created by shifting leaves and grass, as well as human respiratory systems and other body part motion [15–17].

6.3.5 Heartbeat Sensing by Using Doppler Radar

Life signals and respiratory systems can be distinguished using Doppler radar detection technologies. This method can accurately express pulse data and is applicable in the heart rate variability (HRV) evaluation for the casualty's physical and mental condition [18–20]. HRV refers to the change in pulse from beat to beat, which indicates the cardiovascular range and dynamic response of the cardiovascular regulatory frameworks. In the medical field, decreased HRV has been linked to impaired autonomic action, aging, and other factors. HRV research is significant in the context of cardiovascular direction in a variety of illnesses such as inward dying, hypertension, and heart failure. In the current UAD, HRV technology is proposed to assess the percentage of wounds.

6.3.6 Design of Robotic Arm in a UAD

A robotic arm is a form of mechanical arm that is frequently designed on its own. The arm could consist of all the elements or be a part of a larger robot. The connections on the servo controller are linked, allowing rotational and translational movement. The end effector is the final link in the controller's kinematic chain, and it mimics a human hand. Anthropomorphic robot arm [21] was used to evaluate the robotic arms. It's designed to look like a human hand and finger, and it comes with two retractable mechanical arms. Both arms work in a parallel fashion. The UAD will be incorporated with technologies for height determination. Individual arms can be used to identify the correct person, and they will be safely guarded with high precision and positioned in the proper posture.

6.4 UNMANNED AMBULANCE DRONE PROTOTYPE SPECIFICATION DETAILS

Based on the design, the prototype of UAD was fabricated. The prototype is fabricated with the following components:

- Frame (consist of power distribution board),
- Micro controller (KK 2.15 controller board) - 1,
- ESC (30 Amps) (4 pieces),
- Brush-less motor (1,000 kV) (4 pieces),
- Propellers (2 sets),
- Battery -3,000 mAh (Lithium-Polymer)- 1,
- Transmitter and receiver– 1.

The material used for the manufacturing of the robotic arm is glass- fibre composite. The glass-fibre composite required for the robotic arm is manufactured by vacuum bag molding process depicted in Figures 6.4–6.5. Through this process, a

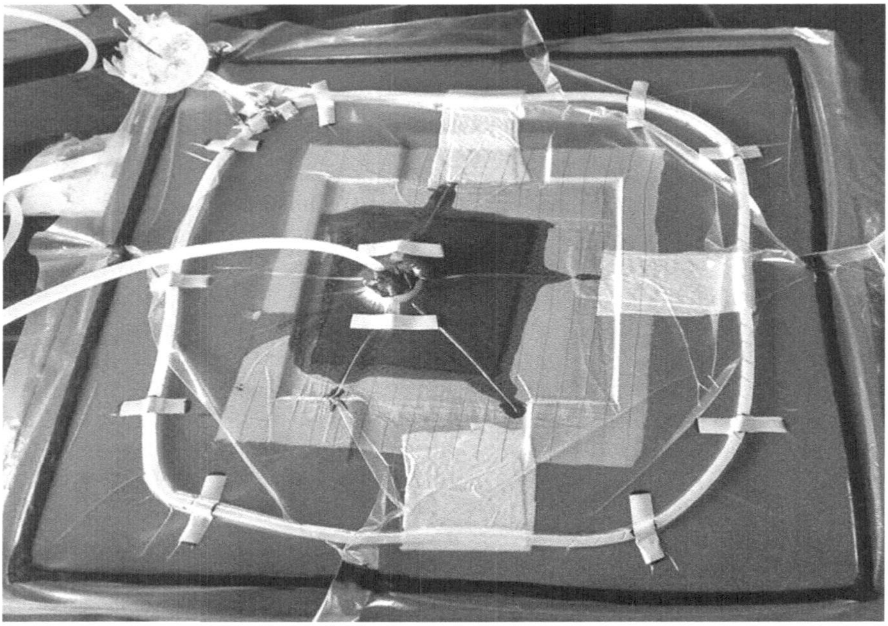

FIGURE 6.4 Composite manufacturing using vacuum mould setup.

FIGURE 6.5 Robotic arm made of glass-fibre composites.

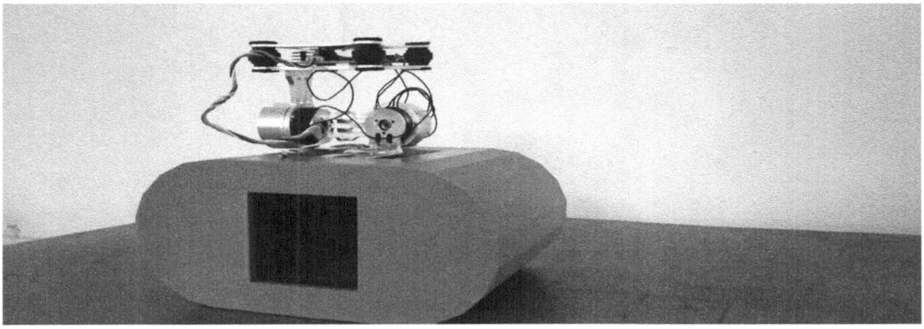

FIGURE 6.6 Tilt mechanism for stability assembled on 3D-printed fuselage.

bidirectional glass-fibre composite is obtained. The fuselage is an elliptical hollow cylinder as it produces less drag on the model and the amount of stress it can withstand is more an elliptical model as there are no sharp edges where failure can happen because of the energy storage over the sharp points. The fuselage model made in solid works software is converted into STL file format and then it is fed to the computer which is used to operate the 3D printer. Then the 3D model of the fuselage is printed. Figure 6.6 shows an arrangement that reduces the airsickness caused to the patients inside the ambulance drone (a two axial gimbal system is consolidated inside the drone). The system consists of two motors to create the required torque (moment), an accelerometer for stabilization of the fuselage. During the manoeuvre, the tilting of the drone takes place, leading to the entire drift of the vehicle as well as of the fuselage. As an example, for the forward movement, the drone must incline its propellers forward to create thrust. A patient in the fuselage may experience these tilt motions and may feel nausea, which leads to airsickness. Accelerometer calculates the inclination throughout the manoeuvre and creates torque to sustain the flat position of the fuselage with the help of motors. Wheels are incorporated into the fuselage, allowing the drone to pass through narrow areas like in the forest, congested areas (Figure 6.7). The placement of the wheels is below the

FIGURE 6.7 Wheels to access small passages or dense forest.

FIGURE 6.8 Prototype of unmanned ambulance drone (UAD).

fuselage leading to ground clearance. Thus, the structure of a UAD prototype is designed and fabricated (Figure 6.8). Thus, a discussion on prototype development of ambulance drones for rescue is conducted in the present chapter.

6.5 CONCLUSION

The various systems needed for a UAD are reviewed and the prototype development from conceptual design stage is discussed in the present chapter. A UAD is capable of VTOL in any terrain condition for emergency, search, rescue, and medical emergency for people. The UAD had been designed in this manner in that it is able to have compact takeoff and touchdown close to the sphere of the affected person, reach the affected person in a brief duration of time, seek and rescue of humans buried beneath earth's rubble after an earthquake, rescue of soldiers/civilians on a conflict field, etc. The drone is proposed with various human detection systems for emergency, search, rescue, and casualty evacuation process in various environmental conditions. Initially, a number of iterations have been done for the aerodynamics weight, structural, propulsion, and performance for finalizing the design. Finally, a model of a UAD is manufactured with a 3D-printed elliptical fuselage, robotic arm fabricated from glass-fibre composites for transferring patients into the fuselage, tilt mechanism to reduce airsickness, and wheels for ground transportation. Thus, a perspective on conceptual design, survey on life detection systems, and prototype development of an ambulance drone for rescue is discussed in the present chapter.

REFERENCES

[1] TU Delft Ambulance Drone, Available online: https://www.tudelft.nl/en/ide/research/research-labs/applied-labs/ambulance-drone (accessed on 20 August 2021).

[2] Health Integrated Rescue Operations, Available online: https://www.hirotelemed.com (accessed on 20 August 2021).

[3] Medicine delivery drone, Available online: https://flyzipline.com (accessed on 20 August 2021).
[4] Hua, M., T. Hamel, P. Morin, and C. Samson. 2013. Introduction to feedback control of under actuated VTOL vehicles: A review of basic control design ideas and principles. *IEEE Control Systems Magazine*, 33(1): 61–75.
[5] Air medical services, Available online: https://en.wikipedia.org/wiki/Air_medical_services (accessed on 20 August 2021).
[6] Zailani, M. A. H., RZAR Sabudin, R. A. Rahman, I. M. Saiboon, A. Ismail, and Z. A. Mahdy. 2020. Drone for medical products transportation in maternal healthcare: A systematic review and framework for future research. *Medicine (Baltimore)*, 99(36): e21967. doi: 10.1097/MD.0000000000021967.
[7] Mohd, S. A., K. B. Gan, and A. K. Ariffin. 2021. Development of medical drone for blood product delivery: A technical assessment. *International Journal of Online & Biomedical Engineering*, 17(9): 183–196.
[8] Rosser, J. B. Jr, B. C. Parker, and V. Vignesh. 2018. Medical applications of drones for disaster relief: A review of the literature. *Surgical Technology International*. 33: 17–22.
[9] Kumar, S., S. Mano, and V. M. Sreehari. 2019. Air ambulance drone for surveillance and casualty evacuation, *International Journal of Mechanical and Production Engineering Research and Development*, 9(4): 1391–1400.
[10] Gardner, L. and T. M. Chan, et al. 2007. Cross-section classification of elliptical hollow sections *Steel and Composite Structures*, 7(3): 185–200.
[11] Nanzer, J. A. and R. L. Rogers. 2007. Human presence detection using millimeter - wave radiometry. *IEEE Transactions on Microwave Theory and Techniques*, 55(12): 2727–2733. doi: 10.1109/TMTT.2007.909872
[12] Li, J., Z. Zeng, J. Sun and F. Liu. 2012. Through-wall detection of human being's movement by UWB radar. *IEEE Geoscience and Remote Sensing Letters*, 9(6): 1079–1083.
[13] Zhong, Y., Z. Zhou, T. Jiang, M. Heimlich, E. Dutkiewicz, and G. Fang. 2016. Classification of animals and people based on radio-sensor network. *16th International Symposium on Communications and Information Technologies*, 113–116. doi: 10.1109/ISCIT.2016.7751603
[14] Setjo, C. H., B. Achmad, and Faridah. 2017. Thermal image human detection using Haar-cascade classifier. *2017 7th International Annual Engineering Seminar (InAES)*, 1–6. doi: 10.1109/INAES.2017.8068554
[15] Bianco, V., P. L. Mazzeo, M. Paturzo, C. Distante, and P. Ferraro. 2020. Deep learning assisted portable IR active imaging sensor spots and identifies live humans through fire, *Optics and Lasers in Engineering*, 124: 105818. 10.1016/j.optlaseng.2019.105818.
[16] Locatelli, M., E. Pugliese, M. Paturzo, V. Bianco, A. Finizio, A. Pelagotti, P. Poggi, L. Miccio, R. Meucci, and P. Ferraro. 2013. Imaging live humans through smoke and flames using far-infrared digital holography. *Optic Express*, 21(5): 5379–5390.
[17] Bianco, V., M. Paturzo, A. Finizio, K. A. Stetson, and P. Ferraro. 2015. Portable IR laser system for real-time display of alive people in fire scenes. *Journal of Display Technology*, 11(10): 834–838.
[18] Li, C., F. Chen, F. Qi, M. Liu, Z. Li, F. Liang, et al. 2016. Searching for survivors through random human-body movement outdoors by continuous-wave radar array. *PLoS ONE*, 11(4): e0152201. 10.1371/journal.pone.0152201
[19] Li, C., F. Chen, J. Jin, H. Lv, S. Li, G. Lu, and J. Wang. 2015. A method for remotely sensing vital signs of human subjects outdoors. *Sensors*, 15: 14830–14844. 10.3390/s150714830

[20] Boric-Lubecke, O., J. Lin, et al. 2008. Battlefield triage life signs detection techniques. *Radar Sensing Technology XII, (67470J) Proceedings of SPIE*, 6947. 10.1117/12.781928.

[21] Del Sol Rodríguez, J., F. López-Colino, G. G. de Rivera, and J. Garrido. 2014. Design and development of an anthropomorphic robotic arm for educational purposes. *Design of Circuits and Integrated Systems, Madrid*, 1–6. doi: 10.1109/DCIS.2014.7035562.

7 A Comprehensive Review on Internet of Agro Drones for Precision Agriculture

P. Pandiyan, Rajasekaran Thangaraj, M. Subramanian, S. Vivekanandan, S. Arivazhagan, and S.K. Geetha

CONTENTS

DOI: 10.1201/9781003252085-7

7.1 INTRODUCTION

The overall investments in the agricultural industry have surged by 80% in the last five years. The investments are meant to accomplish at least 70% growth in 2050 [1], with the aim of fulfilling the demands of a growing population while the area of cultivated land shrinks. In broadacre agricultural activities, the majority of the agricultural farm work is now done by human-driven machines, rather than by hand. Intensive farming methods, which are common in the industry, have resulted in farmers in mechanized agriculture having little hands-on experience with sensing the condition of the field. Remote-sensing approaches have been introduced to assist in precision agriculture by collecting data, which can then be analyzed to track plant growth throughout the cultivation period. During the cultivation, an increasing amount of satellite imaging data has become available. For example, satellite images from the Sentinel-2 satellites are available and distributed by the European Space Agency [2]. Emerging technologies like the unmanned aerial vehicle and the Internet of Things (IoT) can give tremendous promise for smart farming and precision agriculture, which have the potential to bring major benefits by increasing production over the long run [3].

The IoT-based agricultural drone (IAD) opens up new possibilities for precision farming, allowing for actual and site-specific field management [4]. In IAD-based precision agriculture, a system has been developed for observing plants' health status with the goal of managing various key farming processes, such as plant growth monitoring [5], irrigation [6], fertilizer spraying [7], disease detection [8], and so on. In this aspect, cutting-edge technology like the IoT can help with real-time data acquisition from agricultural land. This data may be analyzed and utilized in a timely manner to assist in providing critical decisions in the plant management process [9].

Although unmanned aerial vehicles (UAVs) are a relatively new technology, the very first effort to establish a powered UAV was in 1916 itself [10]. UAVs are also known as drones. In the beginning, drones were originally used for military applications, but they have been rapidly expanded in recent years to a wide range of applications, including commercial, scientific, and agricultural, etc. In the 1980s and 1990s, technological developments and the miniaturization of associated hardware components paved the way for the widespread use of drones. Nowadays, drones are becoming more popular in precision farming with remote-sensing applications [11].

The use of drones equipped with various types of sensors can be utilized to locate which regions of plants require attention to grow effectively. Hence, farmers can respond quickly in the event of a problem being detected. IAD may be utilized for a variety of precision agriculture applications, including plant health monitoring system, growth status monitoring, identifying the disease affected zone, and productivity assessment, weed detection and management [12], and so on. Because drones have become increasingly common in precision agriculture applications in recent years and are widely regarded as the future of remote sensing, it is an area that has attracted great interest.

The rest of this chapter is structured as given below. Section 7.2 provides the scope for precision farming throughout the world. The architecture of the drone in precision agriculture is described in Section 7.3. Section 7.4 of this chapter highlights the significance of drones in precision agriculture and describes the best drones currently available on the market for agricultural monitoring and observation, with the goal of enhancing plant quality and reducing damage to agri-fields. Implementation of IoT architecture in drones is described in Section 7.5. Section 7.6 discusses the applications of drones in precision farming. The difficulties of using drones for precision agriculture are discussed in Section 7.7 and benefits are explained in Section 7.8. A conclusion is drawn with a future scope in the area of precision agriculture using agro drones.

7.2 SCOPE FOR PRECISION AGRICULTURE

Drones in agriculture are already used in Asia, but in other areas of the world, they are only allowed to be used for certain studies such as economic activities in agriculture, forestry, and horticulture [13]. Insect, disease and weed control, the dispersal of fertilizers and micro-granular pesticides, and even new forest planting are just a few of the tasks that drones are increasingly being employed for. The commercial usage of agro drones have been subsidized by the Chinese government [14]. Therefore, DJI private limited educated over 10,000 people to work the Agras MG-1 series 8-rotor sprayer, a 2015 model drone [15]. As demand for multi-rotor drones grew in Japan, Yamaha Motor, which had traditionally focused on unmanned helicopters, began offering them. The spraying quality of the 10-L capacity YMR-08 model, which utilizes co-axial rotors, is equivalent to the Fazer and RMax helicopters in tiny paddy fields where helicopters are ineffective. Yamaha's investigation indicates that operators rent 2,500 or more manual radio-controlled or completely autonomous helicopters to spray roughly 42% of the country's rice fields [16]. In South Korea, around 100 manual RC/fully automated helicopters are in operation, and Yamaha has newly begun to roll out its technology in New Zealand and Australia, as well as in the United States of America, where the Federal Aviation Administration (FAA) has granted Yamaha permission to conduct commercial trial services and research [17]. Another possible use is spraying pesticides to reduce the presence of undesirable plants and weeds in underdeveloped or remote areas, like the Bitou Bush Control Programme (BBCP) of the Great Lakes Council in Australia, which is now being handled by Yamaha's Key Aerial Services. Bracken is a genus of large, coarse ferns in the Dennstaedtiaceae family that thrives in tough places in the United Kingdom's uplands [18,19]. This is mainly a sheep-producing area where the government has allowed a coalition of agricultural drone owners and other parties to test spraying from drones.

Drones are also employed in Denmark to distribute pest-eating beneficial insects to farm crops. A study from South Denmark University notes that the use of certain insects in high-value production within controlled environments is quite manageable. However, employing bio-control in large, open regions has proven to be difficult and expensive in the past, but drones are now making it practical. An Australian drone company, Rise Above, has developed a drone that can dispense seed and fertilizer by having an electric engine and hopper extension.

In India, 40 drone start-ups are working to improve technological standards and lower the cost of agriculture drones so that they are more accessible to farmers [20]. For the past several years, the state government of Maharashtra has encouraged drone firms to work with them. The World Economic Forum's Centre's (WEFC) fourth Industrial Revolution has agreed to examine the usage of drones in various government programs. This was done in cooperation with the Maharashtra government. Agronomists in Maharashtra's Dahanu-Palghar tribal areas have mastered the use of drones for fish farming, bio-control, crop rotation, organic farming, biowaste management, and hydroponics [21].

Operational regulation, the cost of drones, and the scarcity of skilled pilots are all stumbling blocks to the growth of the drone sector in everywhere. Urban drone teams are extremely expensive to recruit for small-scale evaluations and crop planning since they must examine a remote, smaller field. The same way farmers are encouraged to hire farm tools, farmers should be taught in drone operation and encouraged to acquire their own drones. Medium and small-scale farmers are hesitant to employ drones due to the price involved in obtaining them. Aside from cost and technological know-how, a scarcity of trained pilots is a major impediment to the growth of the UAV business in India.

7.3 THE ARCHITECTURE OF DRONES IN PRECISION AGRICULTURE

A drone's basic design comprises of the following components, excluding payload sensors [22] are frame, electronic speed control (ESC) modules, brushless motors, control board, and transmitter, receiver and inertial navigation system (INS). Typically, the drones used in PA consists of the following important components [23].

7.3.1 Mode of Operation

Drones used in precision agriculture are semi-autonomous. The drone must then follow a flight route that is illustrated by waypoints and altitude. Thus, the drone must have a system for measuring its position. (e.g., GNSS- "Global Navigation Satellite System") onboard in order to know its position in relation to the waypoints. It also includes an altimeter (such as a ultrasonic sensor, laser altimeter, or barometer,) for maintaining constant flight altitudes. The software used to plot the mission's path are ArduPilot, PX4, ready-to-fly, Dronecode, Flone, OpenDroneMap, DronePan, Paparazi UAV, MAVLink, BetaFlight, CleanFlight, LibrePilot, APM Planner, and so on.

7.3.2 Ground Control Station (GCS)

GCS can either link to the drone or perform various operations and monitoring on the drone. It is responsible of tracking information related to the drone's flight. When it comes to receiving data, the user has the ability to obtain information that's pertinent to the aircraft's flight path, as well as data acquired by sensor systems used for flying assistance. Furthermore, the GCS comprises the software needed for data handling and the extraction of information that can be viewed by farm managers using the system.

7.3.3 Drone Control System

This system is responsible for controlling the drone. There are two types of data links: a built-in computer and a remote control (two-way data link). This system is comprised of the autopilot system and the flight control system and/or both of which are responsible for controlling the operation of the drone on the ground. During flight, this drone control system collects and processes data obtained from the flight control system or the autopilot system, ensuring that the drone runs properly at all times. It has sensors that keep track of how it flies, including monitoring things like air pressure and distance from the ground. Furthermore, the control system is capable of processing data from sensors in order to fix any issues that may develop, as well as connecting with the GCS electronically and on a real-time basis by receiving and transmitting the required information.

7.3.4 Data Acquisition Sensors

The drone's payload is comprised only of actuators and sensors that are not required for controlling its flight (e.g., the camera attached to the gimbal). Drones are equipped with specialised sensors which evolves into strong sensing systems that supplement IoT-based approaches. The sensors' primary function is to collect high-resolution (HD) images that may be utilized to monitor various plant parameters. A range of various sensor types can be utilised in an agro drone based on the crop metrics that need to be monitored [24]. In contrast, the requirement for limited payload capacity as well as the employment of small platforms impose a number of restrictions on the sensor(s) selection that can be employed. The major requirements for the sensors are small size, low-energy consumption, and low weight. In addition, the chosen sensors should have the ability to take high-resolution images, which is absolutely needed.

Commercial on-board sensors that meet the aforementioned requirements and are utilized for PA primarily fall into five categories: multispectral cameras, RGB cameras, hyperspectral cameras and thermal cameras, LiDAR systems. The different features of the vegetation can be monitored depending on the sensor type, such as colour and texture, or the crops' geometrical shape. Aside from that, few sensors are capable of measuring radiation at specific wavelengths. These sensors are capable of retrieving even more information, such as the amount of plant biomass, the overall health of plants, and the moisture content of the soil.

7.3.4.1 RGB Colour Sensors

RGB sensors [4,25] are commonly employed sensors by drone for precision farming applications. High-quality images can be taken at a lower cost with this type of camera than with other options. In addition, they're easy to use and operate, as well as light weight. Furthermore, the information obtained necessitates straightforward processing. The images can be taken in a variety of weather situations, including both cloudy and sunny days, but a precise time period depending on meteorological conditions is required to avoid the image being over or under-exposed. The fundamental drawback of these sensors is insufficient for monitoring a range of vegetative elements that need non-visible spectral data. These sensors are frequently used in conjunction with others.

7.3.4.2 Multispectral and Hyperspectral Cameras [26,27]

Drone operators can acquire insight into plants by sending their unmanned aircrafts on missions where imaging devices are equipped with multispectral or hyperspectral sensors, as these devices capture data across a broad spectrum of wavelengths. These sensors measure the water capacity in the leaves, total chlorophyll, the NDVI, and the Leaf Area Index to identify the status of the measured vegetation. Spectral data can be quite suitable in assessing a extensive range of physical and biological properties of crops. Chlorophyll absorbs the red channel of visible light, while near infrared (NIR) radiation is largely reflected, allowing sick areas of the crops to be recognized in an image. Even if it isn't detectable in the red channel yet, it can still be seen in the NIR channel. As a result of the application of spectral information, it is possible to calculate a variety of vegetation indices and monitor a variety of agricultural parameters.

Although multispectral and hyperspectral sensors are more expensive than conventional sensors, they are routinely utilized. Advanced pre-processing algorithms are necessary, however, to get the most out of the images collected. Image enhancement, geometric correction, image fusion, and radiometric calibration are frequently included in the pre-processing of spectrum images. As a result, hyperspectral and multispectral sensors differ greatly in the number and width of bands that each sensor is capable of recording. Each type of sensor captures a certain number of bands that is much larger than that of the other sensor types. On the multispectral end of the spectrum, sensors can capture around 5–12 bands; in contrast, on the hyperspectral end, they can capture around 100 or 1,000 bands but have a narrow bandwidth compared to multispectral sensors. In recent investigations, multispectral sensors have been used significantly more than hyperspectral sensors due to its less cost. On the other hand, as a promising new tool for plant phenotyping research, hyperspectral technology is predicted to take off [24,28,29].

7.3.4.3 Thermal Infrared Sensors

Temperature-sensitive thermal cameras have proven a significant potential for detecting water stress in plants, which is caused by an increase in the temperature of the stressed vegetation. Thermal infrared sensors collect temperature data from objects and produce images that show them based on available data rather than apparent features. An infrared sensor and optical lens are used in thermal cameras to gather infrared (IR) energy from the environment. Infrared radiation is emitted by all objects that are warmer than absolute zero at specified wavelengths that are proportional to their temperature. Therefore, thermal cameras concentrate on and detect radiation at these wavelengths, which they subsequently transform to a gray-scale image for the purpose of representing heat. Colour images may be generated by many thermal imaging sensors. Warmer objects are frequently depicted as yellow, while colder objects are depicted as blue in the captured image. This sort of sensor is only employed in an irrigation management. Therefore, they are rarely used in precision agriculture applications of drone systems that are primarily concerned with monitoring other agricultural metrics, such as yield.

RGB sensors are frequently modified to collect data on radiation in other bands as well, most commonly the red edge (RE) or near infrared (NIR) bands. This method is employed when stakeholders wish to avoid incurring the higher costs associated with multispectral camera acquisition. This is accomplished by substituting a NIR perceptible optical filter for original optical filter, frequently resulting in a hybrid version of NIR-RGB sensor. Another embedded RGB sensor will frequently gather the viewable channel. In many instances, both visual and multispectral sensors were used.

It is common practise to use LiDAR systems [30] and RGB cameras to scan the terrain surface in order to develop a “Digital Surface Model” (DSM) or a “Digital Terrain Model” (DTM) of the monitored area. The DTM depicts the soil surface not taking into consideration the height of the plants. On the other hand, the DSM denotes the entire Earth’s surface as well as all of its inhabitants. The DSM and DTM were extrapolated using commercial mapping software [28]. By subtracting the DTM from the DSM, it is possible to generate a differential model of the trailing plant rows.

The following capabilities are required to employ a drone for PA purposes [31]. The drone

a. flies in accordance with the waypoint’s definition,
b. controls its flight altitude,
c. detects and dodge hindrances during flight,
d. lands in accordance with the battery status automatically; and
e. acquires images by stabilizing the gimbal.

The cameras used in agricultural drones are listed in Table 7.1.

TABLE 7.1
Types of Cameras Used in IoT-Based Agro Drones

Type of Camera	Image	Specifications
RGB camera [32]		• Carrier Drone: DJI Phantoms • Colour space/spectral band: sRGB • Focal length: 0.147 cm • Sensor: DJI FC6310
Multispectral camera [33]		• Carrier Drone: DJI M100 • Colour space/spectral band: 470–850 nm • Focal length: 20 mm • Sensor: Slant range 3p/4p/4p+
Hyperspectral camera [34]		• Carrier Drone: DJI M600 • Colour space/spectral band: 400–1000 nm • Focal length: 8 mm • Sensor: SPECIM AFX10 Headwall
Thermal Camera [35]		• Carrier Drone: S800 EVO, Modified Hexacopter • Colour space/spectral band: 7.5–13.5 nm • Focal length: 14,000 mm • Sensor: FLIR Tau 2
LiDAR [36]	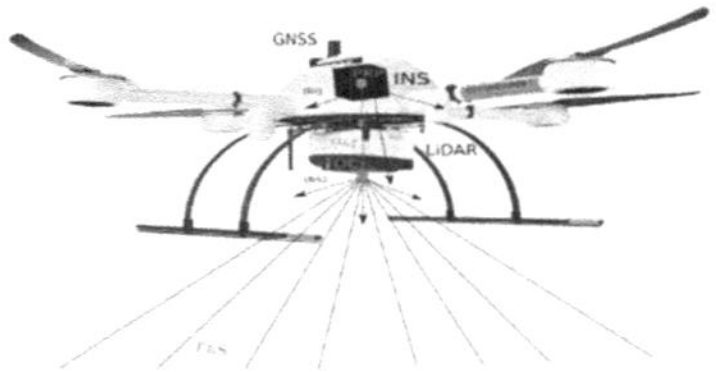	• Carrier Drone: DJI M200 series, LiAir 200 • Colour space/spectral band: 40 channels • Focal length: 100 mm • Sensor: Hesai Pandar40

7.4 AGRO DRONES AVAILABLE ON THE MARKET FOR PRECISION AGRICULTURE

The modern agricultural industry is at a crossroads. Industry professionals benefit greatly from recent advancements in precision agriculture techniques because of the increase in modern management strategies available. For the development of standard operating procedures and artificial intelligence algorithms, the utilization of several types of agro drones for specific plant-growing operations is being investigated, specifically for the realization of electronic maps of fields, reviewing the germination of seeds, making useful observations of plant growth conditions, assessing ploughing quality, assessing crop yields, and maintaining environmental monitoring of agricultural land.

The drone market for agricultural needs has grown significantly during the last few years. Drones with spray tank add-ons can cost anywhere between 3 and 7 lakhs, based on the spray tank's make and capacity. The companies that have registered for drone manufacturing are tabluated in Table 7.2 [21].

TABLE 7.2
Agro Drones Available in the Market [21]

Manufacturer name	Make name	Applications
UAVE Limited	Prion Mk3	• Survey planning, • Payload integration and • Data acquisition
Detect Technologies	NOCTUA DS	• Anomaly detection in real-time environment • Optimize the industrial productivity
Hubblefly Technologies	Starlite	• Specifically, for high to low-level visibility inspections and surveillance • Meet the training requirements for unmanned aerial systems
Throttle Aerospace Systems	LookOut VTOL	• Multiple sensors, including as RGB, EO/IR, multispectral, and radiometric can be swapped out on-the-fly in the field.
Skylark Drones Pvt Ltd	Patang	• Transforming the drone ecosystem from compliance to fleet management to worksite intelligence
ideaForge Technology Pvt Ltd	Ninja	• Defence and homeland security • Renewable energy monitoring and mining.
Asteria Aerospace Pvt Ltd	A400	• Defence and homeland security • Renewable energy monitoring and mining.
Aarav Unmanned Systems Pvt Ltd	Insight 2.0	• Surveying and mapping
DJI Agriculture	P4 Multispectral, AGRAS T30, MG-1P, Phantom 4 RTK, DJI Terra	• Provides intelligent solutions • Farmers and growers, as well as agriculture and service sectors, all stand to benefit from this product.

(Continued)

TABLE 7.2 (Continued)
Agro Drones Available in the Market [21]

Manufacturer name	Make name	Applications
senseFly	eBee X, eBee Geo, eBee Ag	• Surveying and mapping • Mining, quarries, and aggregates • Engineering and construction • Environmental monitoring • Humanitarian
PrecisionHawk	DJI MATRICE 200V2, DJI phantom series	• Collect, process, analyze, deliver, and action for smart farming
Parrot	ANAFI, ANAFI Thermal, ANAFI FPV, ANAFI Extended.	• Observe the environment from visible to invisible and complete mission immersion
AeroVironment	Puma LE, Raven B, Wasp AE, JUMP 20, T-20	• Assisting individuals in acquiring actionable intelligence that enables them to confidently navigate their way into a more safe, secure, and successful future.
Microdrones	mdLiDAR3000LR, and mdLiDAR1000HR	• Surveying • Environmental monitoring • Infrastructure inspection • Precision farming
AgEagle	AgEagle RX-60, AgEagle RX-48	• Provide the data, tools, and strategies required to create and implement drone-enabled solutions that solve critical problems
American Robotics, Inc.	American Robotics Scout drones	• Scout systems to collect data on a regular basis or to launch missions as needed

7.5 IMPLEMENTATION OF IOT ARCHITECTURE

Figure 7.1 depicts a high-level overview of the system to detect the disease in the agricultural field and find out the disease affected position accurately [37]. Sensor devices, gateway, M2M services, and application are the four components of the system that are implemented according to IoT architecture. Drones, cameras, and GPS sensors are examples of sensor devices. First and foremost, the drone is tasked with monitoring a large area using GPS to collect location data in the form of latitude and longitude. The image of the plant leaf on the field is then captured with a camera. The camera is a high-definition camera capable of capturing RGB colour images in JPEG format. The camera and GPS communicate with each other and transfer data to the ground control centre through a WiFi device. Drone flying and image capture are controlled manually by staff, while data is delivered automatically and instantaneously via WiFi.

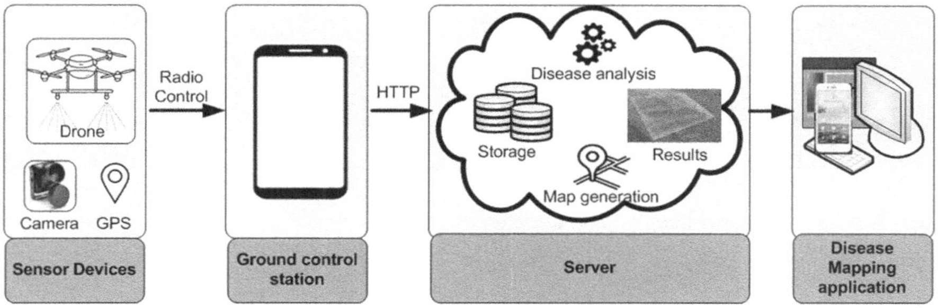

FIGURE 7.1 Concept diagram of plant disease detection and mapping system.

Three functions are performed by the ground control station. First, it directs the drone's flight to gather images from specific locations. Second, it synchronizes and saves all of the images captured by the camera. Third, it is a client that uses the HTTP protocol to send data to the server for additional processing. A ground control station can be a mobile or tablet application platform that provides convenience to users.

The server is where data computation and analysis take place. In this system, a web server application that uses the HTTP protocol is recommended. For starters, it functions as a data storage device that is classified into two categories: field plant images and its coordinates. As a result of this, it performs disease analysis on plant leaves using images captured, and then applies the results to the image processing stage. The flow of image processing is depicted in Figure 7.2 as five processes: image collection, pre-processing, segmentation, feature extraction, and classification. It also creates a map of agricultural fields. Following the creation of the map and the analysis of the disease, the system will deliver the results in the form of a map that depicts the location of disease occurrence on the map that was made. The application runs on a smartphone and displays disease occurrences on a agricultural field map for the user to check for further analysis. The implementation of IoT architecture is illustrated in Figure 7.3.

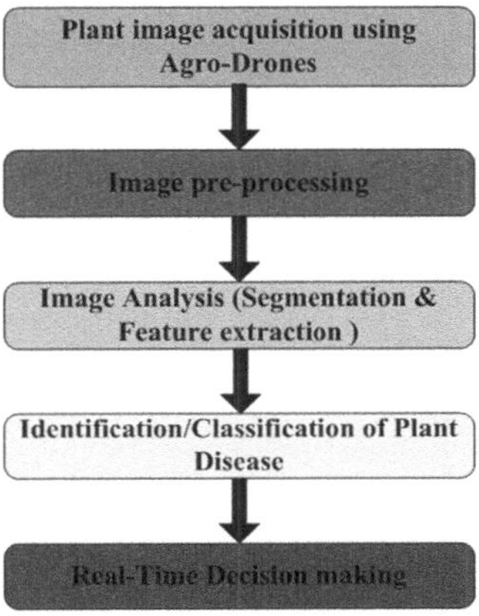

FIGURE 7.2 Image processing operation framework.

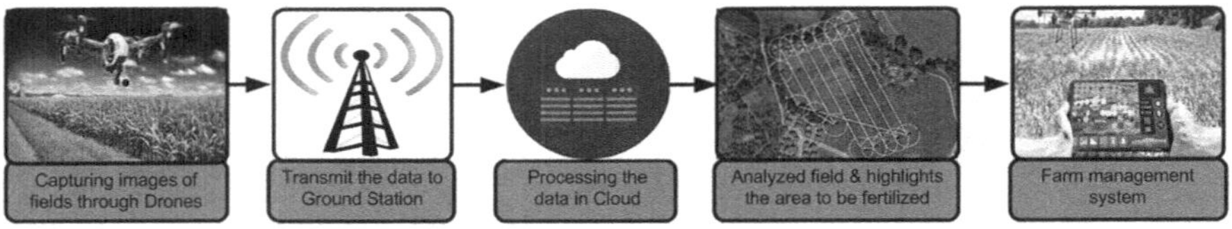

FIGURE 7.3 Implementation of IoT architecture.

7.5.1 Image Processing Techniques Implementation

7.5.1.1 Image Collection

In precision agriculture, a drone is utilized to patrol the agricultural fields to collect images. The image data from the drone's linked camera will be in RGB written colour. Each image has a coordinate tagged on it that can be used for the next mapping phase. The images depict the plant leaves in the various areas of the farm that were examined.

7.5.1.2 Image Pre-processing

The image enhancement of a plant leaf in order to emphasise the texture and disease colour by sharpening the image colours or intensities to better the outcome of future operations. This is advantageous for segmentation procedures that make use of the colour intensity as a parameter. Pre-processing is used to reduce image size and memory use in the processing system by performing image downsizing. To improve contrast, histogram equalisation was employed to modify image intensities. As a result, a clearer sight of the leaf's edge and any infections that have occurred can be seen.

7.5.1.3 Image Segmentation

To isolate the image intensity from the colour information, the hue, saturation, value (HSV) colour space was employed. Figure 7.4 depicts how the colours in the

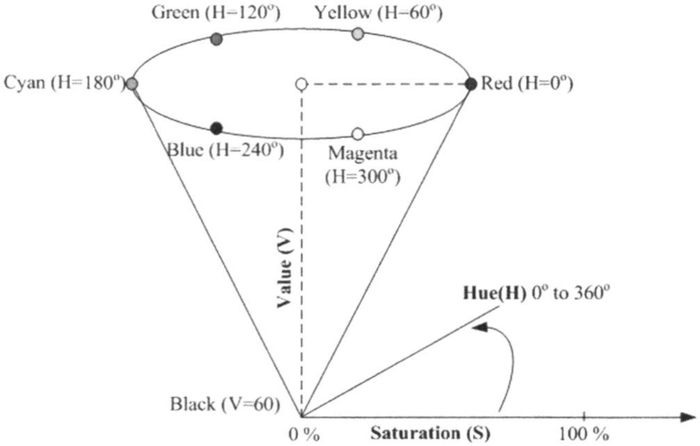

FIGURE 7.4 Hue-saturation-value model.

HSV model are described. Hue is measured in degrees ranging from 0 to 360. Saturation describes the colour's depth, whereas value depicts the colour's lightness as it progresses from black to white. Using a colour separation technique to distinguish between the disease element of the leaf image and the rest of the leaf image's colour (which was mostly green), diseases were detected.

7.5.1.4 Extracting Features

Disease sections' features are separated into two types: shape and colour features. The colour and form features will be employed in classification because each type of disease has various colour and shape properties. For the colour features, the size of the minor and major axes of the disease are compared, and for the form features, the colour's hue from HSV space is determined.

7.5.1.5 Classification

Classification is the process of analysing image features and categorising them into distinct groups. The AI model is employed in order to classify the images.

7.5.2 Map Generation Technique

A precise agricultural field map will be built using GPS data from the drone. The GPS data revealed the longitude and latitude coordinates. After constructing the map, it will be used to map the data collected about diseases that have been investigated. The system uses GPS data linked with the original image to show the location of diseases on an agriculture field map, and the sign differentiates which diseases are present in that location.

7.6 APPLICATIONS OF AGRO DRONES IN PRECISION AGRICULTURE

Currently, drones have found their most successful application in PA, like the use of herbicides only where they are needed, the detection of water deficiencies, and in the investigation of diseases. It is possible to make several decisions using data collected by the drones, such as resolving any detected issues or optimize the reaping by assessing the yield. The applications of agro drones in precision agriculture is illustrated in Figure 7.5. According to published literature, the most common uses of drones for PA are as follows [38].

7.6.1 Weed Mapping and Management

Weed mapping [39] is one of the most widely used applications of unmanned aerial vehicles (UAVs) in PA. Plants that cause issues in agricultural crops called weeds. Plants develop at a slower rate because they are competing for finite resources like water and space. This results in reduced food yields and slower plant growth. Aside from interfering with the growth of crops, weeds can also pose a problem when harvesting is in progress. Herbicides are the most frequently used method of weed control. The most frequent farming practise for controlling weeds involves spraying

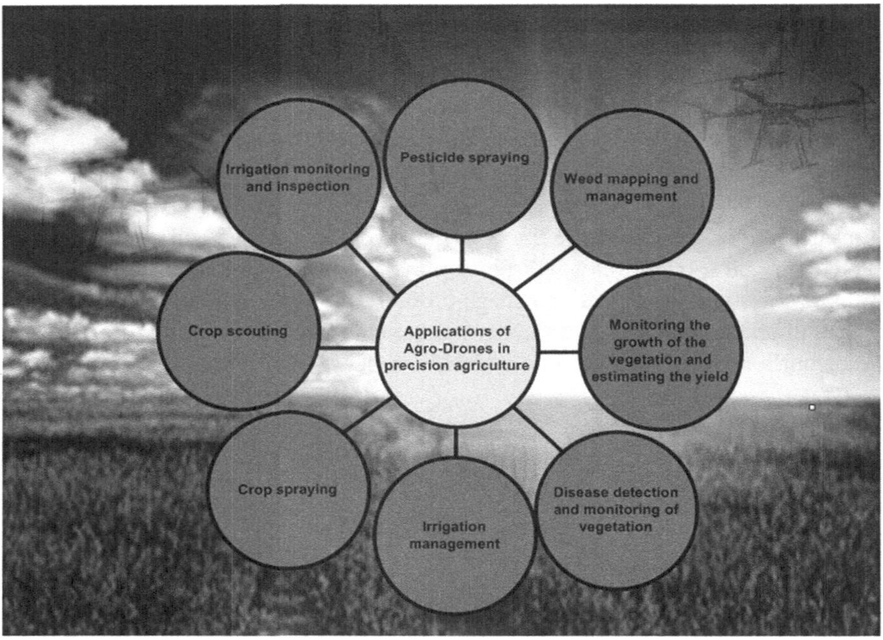

FIGURE 7.5 Applications of agro drones in precision agriculture.

the same doses of herbicides all throughout the field, even when there are no weeds. On the other side, excessive pesticide use might result in the emergence of herbicide-resistant weeds that harm crop output. Additionally, it is a substantial source of pollution in the environment. Furthermore, the previously indicated practise has a major monetary impact. PA practises apply site-specific weed management (SSWM) to address the aforementioned challenges. Instead of spraying herbicides across the entire field, SSWM refers to the application of herbicides in a spatially diversified manner. Weed plants only spread to a few regions of the field; hence, this type of crop is treated using management zones to break up the field into manageable sections. In order to spray herbicide with precision, an accurate weed cover map must be created. The images and data gathered by unmanned aerial vehicles (UAVs) in the field are useful in the creation of an accurate map of the weed cover, which reveals the places where pesticides are required the most, the most infrequently, and the least frequently.

7.6.2 Monitoring the Growth of the Vegetation and Estimating the Yield

Other applications of drones include monitoring vegetation growth and estimating yield [40]. One of the greatest impediments to boosting agricultural output and quality is the lack of tools for routinely assessing the cultivation. These problems are made more difficult due to weather conditions that can change crop microclimates, making it difficult for farmers to produce adequate food. With the regular gathering of information and visualization of plants using drones, there are more options to monitor crop growth and allows for more accurate measurements of several field parameters. Numerous recent studies have concentrated on monitoring the nitrogen status and biomass of crops, as well as yield estimation. The most widely measured crop characteristic connected with nutrient content is biomass, which can be used to assess whether extra fertilizer or other activities are required. Additionally, the information collected by drones can be used to create 3D digital maps of the plants, as well as to measure a variety of parameters like distance between plants or between rows, crop height, and the LAI, among other things. Drones have the ability to systematically collect plant data, allowing farmers to plan plant management, input utilization (e.g., fertilizers), harvesting schedule, and soil and yield diseases in a controlled manner, as well as identify any management errors.

7.6.3 Disease Detection and Monitoring of Vegetation

UAVs are also employed to monitor the health of the plants [41]. Crop health is a critical element that must be checked, as illnesses in crops can result in large financial losses owing to lower yields and lower quality. Crops should be closely checked at all times to discover diseases early and prevent them from spreading. Traditionally, this duty has been carried out by human experts on the ground. This, however, can be very intensive, as inspecting a whole crop can take months, preventing the benefits of "continuous" monitoring. Pesticide use in specific dates is another typical disease control approach. This approach is costly, and it also increases the risk of pesticide residues in the products contaminating groundwater. Disease control is done on a site-by-site basis in precision agriculture. PA practises use a decision-based disease management technique that heavily relies on automated non-destructive crop disease detection. Disease detection is possible because infections cause changes in the crops' biophysical and biochemical features. Crop imaging data is used by UAV-based data processing tools to assess changes in plant biomass and health. As a result, infections can be recognised early on, allowing farmers to respond to minimise losses. This context provides a useful example for using UAVs, in two stages: (a) early detection of an infection, in which UAVs are able to collect health-relevant information to map a potential infection and pinpoint it to the disease's size and location; and (b) treating the infection by using UAVs to target and treat infected locations.

7.6.4 Irrigation Management

UAV technologies have the potential to revolutionise crop irrigation management in PA. Irrigating crops consumes 70% of all water consumed globally [42,43], emphasising the necessity of precision irrigation approaches. Precision irrigation techniques can boost water efficiency by ensuring that the resource is used successfully in the following ways: (a) at the right places, (b) at the right times, and (c) in the right amount. Identifying areas that require extensive irrigation can assist farmers in conserving time and water resources. Simultaneously, these precision farming techniques may result in an increase in crop productivity and quality. Precision agriculture divides the field into distinct irrigation zones in order to manage resources precisely. The use of unmanned aerial vehicles equipped with the appropriate sensor types enables the identification of areas of a crop that require additional water. Simultaneously, the aforementioned technologies enable the creation of specialized maps that depict the soil's morphology, thereby facilitating more efficient irrigation planning for each crop separately.

7.6.5 Corps Spraying

Crop spraying [44] is a precision agriculture application of unmanned aerial vehicles (UAVs) that is less commonly encountered. Manual air-pressure and battery-powered knapsack sprayers are the major spraying equipment used in conventional farming. On the other hand, conventional sprayers have the potential to cause large pesticide losses. Additionally, operators must be present while spraying, resulting in operator exposure. Additionally, spraying the entire field may be time consuming, limiting not only resources but also resulting in insufficient spraying. UAVs can help in this situation since they limit operator exposure and allow for timelier and spatially resolved chemical application. UAVs can follow the contour of the ground while maintaining a steady height thanks to the employment of sophisticated distance measuring devices. As a result, an aircraft can spray the right quantity of herbicide in the right place at the right time, altering both its height and the amount sprayed to match the crop site. Crop spraying is critical in situations where diseases have been identified and pesticide use must be reduced without compromising crop yield. In conclusion, UAV-based systems have the potential to significantly improve crop spray management.

7.6.6 Crop Scouting [45]

A drone can be flown over a field to view, capture, and playback colour images or high-definition video, which can provide a great deal of useful data. It is no longer necessary for growers to drive around the entire property in search of trouble spots. After doing so, they can then travel to that location and conduct additional research.

7.6.7 Irrigation Monitoring and Inspection

In the event of irrigation system failures, locating and repairing them becomes difficult and time-consuming for growers who have multiple fields located across a region [46]. Amid-season inspections of irrigation systems can be cumbersome once crops like corn reach certain heights, forcing inspectors to wade through the crops in order to find the problematic ones. Even though this is a very time-consuming task that can be accomplished with a small, off-the-shelf prosumer camera drone, it is best accomplished with professional drones that are equipped with camera zoom functionality.

7.6.8 Pesticide Spraying

The remote-control drones can be used to spray liquid herbicides, pesticides, and fertilizers [47] on agricultural fields. The use of drones in agriculture has been addressed by the American University of Vertebrate Science as "a safer and more cost-effective way to manage crops." However, the reality is that they do not hold a lot of liquid (only 10–20 kg). For the most part, this is enough, where the use of pesticides is limited to spraying on crops/plants in small allotments on hilly terrain.

Many other tasks besides typical functions of UAVs have been identified, including analyses of soil chemical composition [48,49], the selection of cotton genotypes [50], and detection of mammals [51].

7.7 CHALLENGES TO IMPLEMENTING PRECISION AGRICULTURE USING AGRO DRONES

Compared to expectations, the advancement of precision agriculture has been slower than expected. A sophisticated decision-support system is still required to make timely decisions. A lack of attention is limiting the rise of precision agriculture to temporal variance, a lack of whole-farm emphasis, precise crop quality evaluation methodologies, poor product monitoring, and environmental auditing. The following points discuss some of the constraints associated with precision agriculture [38,52] and also illustrated in Figure 7.6:

a. ***Interoperability of various standards:*** As technology advances, scientists are creating new tools and Internet of Things platforms, which will allow them to work together more seamlessly. Their ability to communicate with one another in the future may cause concern. To be successful in the

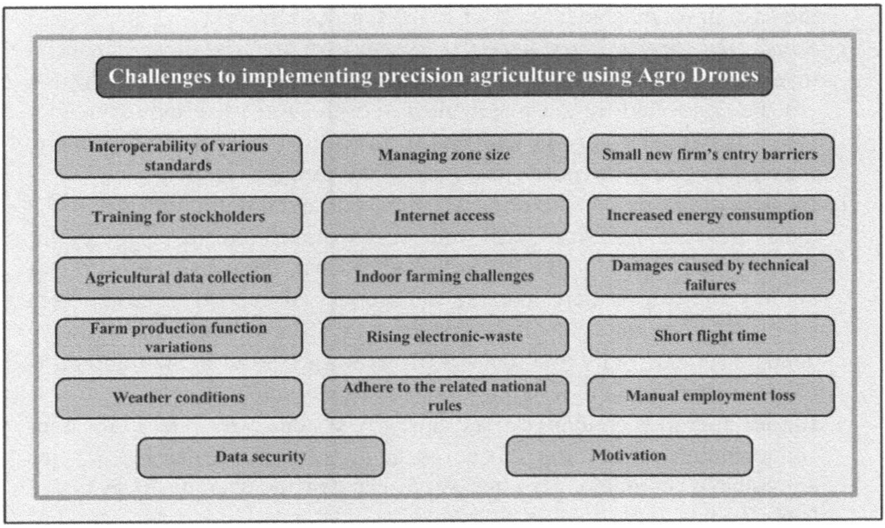

FIGURE 7.6 Challenges to implementing precision agriculture using agro drones.

future, standalone devices and gateways must be integrated into comprehensive, farmer-friendly platforms.

b. ***Training for stockholders:*** Precision agriculture training necessitates the use of cutting-edge technology. In case of small farmers with inadequate knowledge on setting up wireless sensor networks and IoT will be difficult. The importance of farmer training on various precision agriculture instruments is critical, and the effectiveness of PA will be determined by the training. Inadequate knowledge can have a detrimental effect on the quality and yield.
c. ***Internet access:*** There are many rural areas in the world that lack reliable Internet access. PA will continue to be a problem as long as bandwidth speed and network performance do not improve significantly. On the other hand, cloud-based computing requires strong network connectivity. Receiving GPS signals in farmlands with densely populated green environments or hilly terrains can also be one of the significant problems.
d. ***Agricultural data collection:*** Modern PA data is gathered from multiple sources (sensors). On a regular basis, it's impossible to keep track of every sources and manage them throughout the course of a growing season. In fact, the problem becomes more severe when it is realized in large multi-crop farms.
e. ***Farm production function variations:*** It is essential to conduct a full analysis to figure out the appropriate production function i.e., output as a function of key inputs, such as nutrients, fertilizers, irrigation, etc. On the other hand, the production function varies depending on the crop, the plant growth cycle and the farm zone. Without defining the production function correctly, the possibility of application of inputs in incorrect amounts will always exist.
f. ***Managing zone size:*** In the past, farmers thought their entire field as a single farming unit when determining the size of their management zones. This strategy cannot be worked out for implementing PA and leads to increase in establishment cost of IoT setup per farmer.
g. ***Small new firm's entry barriers:*** Because of the significant financial investment required in infrastructure and technology upgrades, the PA will be controlled by the major players in the Agro-IoT industry. As a result, the smaller players will find it difficult to compete in the future. Inadequate competitiveness may exacerbate farmers' burdens.
h. ***Increased energy consumption:*** In precision agriculture, the use of the most efficient resources will aid in the transition to a more environmentally friendly planet. on the other hand, the use of an excessive number of gadgets may increase the amount of energy required. As a result, more resources will be required in order to meet the growing demand for electricity. In order for PA to be successful in the future, it is necessary to focus on designing more energy-efficient gadgets.
i. **Indoor farming challenges:** As most PA systems are suitable for conventional outdoor farming, this raises the question of whether PA systems are suitable for indoor farming. As a result of the scarcity of available land, farmers are becoming more involved in vertical indoor farming. For the time being, PA does not support these indoor agricultural practises.

j. **Damages caused by technical failures:** In the event of a malfunction, agriculture's high reliance on technology could cause serious damage. If a soil moisture sensor fails, a crop may experience water stress. In order to meet this requirement, robust technology and sensors are required to be developed to respond quickly in the event of a malfunction.
k. ***Rising electronic-waste:*** The regular upgradation of hardware components in order to make energy efficient system will result in a buildup of obsolete hardware. These outdated electronic devices and IoT tools could have significant implications in the future. The proper disposal of electronic waste must be planned in advance.
l. ***Short flight time:*** The majority of commercial agro drones have a short flight time, ranging from 20 minutes to one hour, and cover a very small area with each flight. UAVs with a longer flight time are more expensive than those with a shorter flight time.
m. ***Weather conditions:*** The use of agro drones is highly dependent on weather conditions. For example, if it is extremely windy or rainy, the drone should be postponed.
n. ***Adhere to the related national rules:*** The use of agro drones for agricultural purposes is considered commercial, drone flights must comply with all applicable federal, state, and local laws and regulations. In practise, these regulations state that, once a agro drone operator has received the necessary authorization, he or she will be free to operate drone without any restriction.
o. ***Manual employment loss:*** In the future, a substantial number of agricultural workers may lose their jobs. As a result, other industries must be prepared to absorb the workers who have lost their jobs.
p. ***Data security:*** For PA to succeed, data must be well protected against viruses and data theft. As a result, the PA framework should be able to fend off hacking attempts.
q. ***Motivation:*** The beneficial effects of PA take time to manifest. As a result, farmers must be encouraged to participate in PA implementation; otherwise, PA will remain a theoretical concept.

7.8 BENEFITS OF DRONES IN AGRICULTURE

Drone technology is advantageous for a variety of purposes, such as giving simple ways to monitor specific areas of crops and entire fields from a distance. Drones assist farmers in addressing a variety of emerging difficulties in agriculture [21]. The benefits of drones in precision agriculture are illustrated in Figure 7.7.

7.8.1 SAFETY

Drones used for agricultural spraying are operated from a distance by operators. As a result of this process, farmers and farm workers are no longer in direct contact with toxic substances and hazardous operating circumstances.

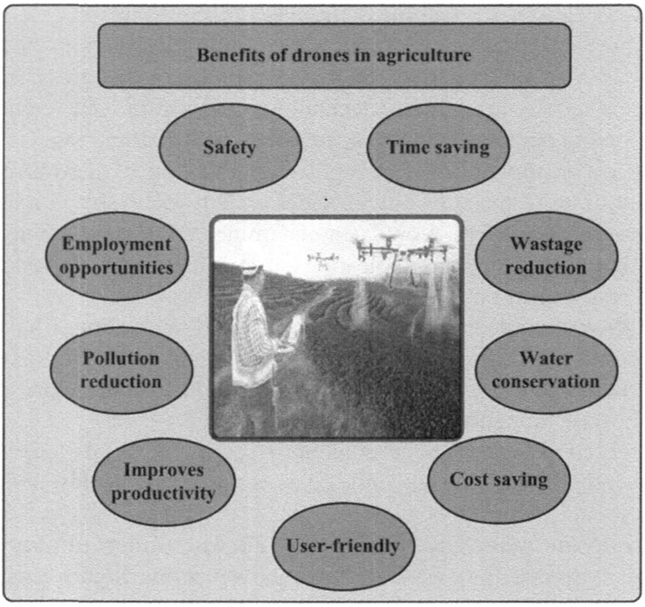

FIGURE 7.7 Benefits of agro drones in precision agriculture.

7.8.2 Time Saving

Drones experience very few delays while on route to the field, as well as less prone to other field operational delays. It is possible to spray between 50 and 100 acres each day with a drone, which is approximately 30 times more than the standard knapsack sprayer.

7.8.3 Wastage Reduction

During pesticide spraying, over 30% of the pesticide can be saved because of the increased automation. Pesticides applied in the form of chemical fog can be applied at all stages of the crop's development.

7.8.4 Water Conservation

Drones use ultra-low volume spraying technology, which saves 90% of the water used in typical spraying methods.

7.8.5 Cost Saving

Drone spraying is significantly less expensive than conventional spraying methods, saving up to 97% on the total cost of spraying.

7.8.6 User-Friendly

The agricultural drone has been designed to be simple to operate and maintain. It offers a minimal maintenance cost, long productive life span, and ease of repair of individual parts as and when they are required.

7.8.7 Improves Productivity

Drones have the potential to satisfy the expanding demands of the population, which is one of its advantages in farming. Using drones to look at crops in the field has helped farmers get around problems. Today's most advanced means of crop monitoring is satellite imaging. However, this technology has a lot of drawbacks, which has prompted farmers to turn to drones instead. Satellite imaging is lacking in precision because images are only obtained once in a day, which is deficient for agronomists. Agricultural drones, on the other hand, can give as many photographs as farmers need in real time, allowing them to be more precise and productive.

7.8.8 Pollution Reduction

Pesticide spraying in the fields can be inconsistent with the help of drones. The usage of agricultural drones can help farmers uniformly spray chemicals to avoid health issues. Drones with the necessary equipment may scan the ground and effectively spray the pesticide wherever it is required, decreases the quantity of pesticide used by farmers and minimizing their expenditures. Finally, it turns out to reduce pollution too.

7.8.9 Employment Opportunities

Drone technology also provides a plethora of employment opportunities for people living in rural areas, staring from computer operator positions to drone machinists. Agriculture is the primary source of revenue for a large section of India's population, and increasing agricultural output increases employment options for them.

7.9 CONCLUSION

New technologies are being employed to increase yield during the previous decade, and precision agriculture is putting them to use. This technology is advantageous in situations when human involvement is impossible, such as spraying chemicals on crops or in areas with a scarcity of labour. Additionally, it simplifies and expedites spraying. This chapter provides an overview of the usage of drones in precision agriculture. The most prevalent applications, the types of agro drones available in the market, and PA's internet of agro drone architecture have all been thoroughly investigated. Next, this chapter gives the details about the most prevalent crop monitoring sensors, as well as the processing methods that leverage agro drones images. Precision agriculture with agro drones is still in its early phases, with plenty of possibility for technological and agricultural advancements. The use of low-cost commercial mini- or micro-drones is predicted to be a future research trend in this field. On the other hand, meeting the measurement precision criteria is challenging, and as a result, a variety of challenges arise.

REFERENCES

[1] Food and Agriculture Organization. (2009). Declaration of the world summit on food security. *WSFS 2009/2, Food and Agriculture Organization of the United Nations, Rome, Italy.*

[2] Drusch, M., Del Bello, U., Carlier, S., Colin, O., Fernandez, V., Gascon, F., ... & Bargellini, P. (2012). Sentinel-2: ESA's optical high-resolution mission for GMES operational services. *Remote sensing of Environment, 120*, 25–36.

[3] Mylonas, P., Voutos, Y., & Sofou, A. (2019). A collaborative pilot platform for data annotation and enrichment in viticulture. *Information, 10*(4), 149.

[4] Esposito, M., Crimaldi, M., Cirillo, V., Sarghini, F., & Maggio, A. (2021). Drone and sensor technology for sustainable weed management: a review. *Chemical and Biological Technologies in Agriculture, 8*(1), 1–11.

[5] Juyal, P., & Sharma, S. (2021,). Crop Growth Monitoring Using Unmanned Aerial Vehicle For Farm Field Management. In *2021 6th International Conference on Communication and Electronics Systems (ICCES)* (pp. 880–884). IEEE.

[6] Sibanda, M., Mutanga, O., Chimonyo, V. G., Clulow, A. D., Shoko, C., Mazvimavi, D., ... & Mabhaudhi, T. (2021).Application of drone technologies in surface water resources monitoring and assessment: A systematic review of progress, challenges, and opportunities in the global south. *Drones, 5*(3), 84.

[7] Aarikar, L., Behaniya, H., Zade, L., Gaikwad, K., Ajmani, S., & Kumar, M. (2021). A review on design and development of agricultural fertilizer spraying drone and multitasking system. *International Journal of Progressive Research in Science and Engineering, 2*(7), 24–28.

[8] Lee, J. M., Lee, Y. H., Choi, N. K., Park, H., & Kim, H. C. (2021). Deep-learning-based plant anomaly detection using a drone. *Journal of the Semiconductor & Display Technology, 20*(1), 94–98.

[9] Saha, H. N., Roy, R., Chakraborty, M., & Sarkar, C. (2021). Development of IoT-based smart security and monitoring devices for agriculture. *Agricultural Informatics: Automation Using the IoT and Machine Learning*, 1, 147–169.

[10] Taylor, J. W. R., & Munson, K. (1977). *Jane's pocket book of remotely piloted vehicles: robot aircraft today*. Collier Books.

[11] Castineira, S., Delwar, T., Duran, R., & Pala, N. (2021). UAV-based agricultural monitoring and data acquisition system for precision farming. In *Sensing for agriculture and food quality and safety XIII* (Vol. 11754, p. 117540G). International Society for Optics and Photonics.

[12] Raja, G., Dev, K., Philips, N. D., Suhaib, S. M., Deepakraj, M., & Ramasamy, R. K. (2021, May). DA-WDGN: Drone-Assisted Weed Detection using GLCM-M features and NDIRT indices. In *IEEE INFOCOM 2021-IEEE Conference on Computer Communications Workshops (INFOCOM WKSHPS)* (pp. 1–6). IEEE.

[13] Yaqot, M., & Menezes, B. C. (2021, August). Unmanned Aerial Vehicle (UAV) in Precision Agriculture: Business Information Technology Towards Farming as a Service. In *2021 1st International Conference on Emerging Smart Technologies and Applications (eSmarTA)* (pp. 1–7). IEEE.

[14] Kendall, H., Clark, B., Li, W., Jin, S., Jones, G., Chen, J., ... & Frewer, L. (2021). Precision agriculture technology adoption: A qualitative study of small-scale commercial "family farms" located in the North China Plain. *Precision Agriculture, 23*, 1–33.

[15] Cheng, L. (2021, July). Product Meaning-Making in High-Tech Companies: A Case Study of DJI Drones. In *International Conference on Human-Computer Interaction* (pp. 329–336). Springer, Cham.
[16] Takao, K., Wataru, K., Mitsuru, U., Takayuki, K., Takao, O., Watanabe, A., ... & Yoshiaki, H. (2021). Aeromagnetic survey in Kusatsu-Shirane volcano, central Japan, by using an unmanned helicopter. *Earth, Planets and Space (Online)*, *24*(1), 971–981. Page range: 1-12.
[17] Mohamed, E. S., Belal, A. A., Abd-Elmabod, S. K., El-Shirbeny, M. A., Gad, A., & Zahran, M. B. (2021). Smart farming for improving agricultural management. *The Egyptian Journal of Remote Sensing and Space Science.*
[18] Lindenmayer, D. B., Wood, J., MacGregor, C., Buckley, Y. M., Dexter, N., Fortescue, M., ... & Catford, J. A. (2015). A long-term experimental case study of the ecological effectiveness and cost effectiveness of invasive plant management in achieving conservation goals: Bitou Bush control in Booderee National Park in Eastern Australia. *PLoS One*, *10*(6), e0128482.
[19] Merkert, R., & Bushell, J. (2020). Managing the drone revolution: A systematic literature review into the current use of airborne drones and future strategic directions for their effective control. *Journal of air transport management*, *89*, 101929.
[20] Gupta, S. K., Kumar, R., Limbalkar, O. M., Palaparthi, D., & Divte, P. R. (2019). Drones for future agriculture. *Agriculture & Food: e-newsletter*, 16.
[21] Pathak, H., Kumar, G. A. K., Mohapatra, S. D.,Gaikwad, B. B., & Rane, J. (2020). Use of Drones in Agriculture: Potentials, Problems and Policy Needs, Publication no. 300, ICAR-NIASM, pp 13+iv.
[22] Brzozowski, B., Daponte, P., De Vito, L., Lamonaca, F., Picariello, F., Pompetti, M., ... & Wojtowicz, K. (2018). A remote-controlled platform for UAS testing. *IEEE Aerospace and Electronic Systems Magazine*, *33*(8), 48–56.
[23] Gupta, S. G., Ghonge, M. M., & Jawandhiya, P. M. (2013). Review of unmanned aircraft system (UAS). *International journal of advanced research in computer engineering & technology (IJARCET)*, *2*(4), 1646–1658.
[24] Yang, G., Liu, J., Zhao, C., Li, Z., Huang, Y., Yu, H., ... & Yang, H. (2017). Unmanned aerial vehicle remote sensing for field-based crop phenotyping: Current status and perspectives. *Frontiers in plant science*, *8*, 1111.
[25] Huang, Y., Reddy, K. N., Fletcher, R. S., & Pennington, D. (2018). UAV low-altitude remote sensing for precision weed management. *Weed Technology*, *32*(1), 2–6.
[26] Honrado, J. L. E., Solpico, D. B., Favila, C. M., Tongson, E., Tangonan, G. L., & Libatique, N. J. (2017, October). UAV imaging with low-cost multispectral imaging system for precision agriculture applications. In *2017 IEEE Global Humanitarian Technology Conference (GHTC)* (pp. 1–7). IEEE.
[27] Singh, K. D., & Nansen, C. (2017, August). Advanced calibration to improve robustness of drone-acquired hyperspectral remote sensing data. In *2017 6th International Conference on Agro-Geoinformatics* (pp. 1–6). IEEE.
[28] Mogili, U. R., & Deepak, B. B. V. L. (2018). Review on application of drone systems in precision agriculture. *Procedia computer science*, *133*, 502–509.
[29] Puri, V., Nayyar, A., & Raja, L. (2017). Agriculture drones: A modern breakthrough in precision agriculture. *Journal of Statistics and Management Systems*, *20*(4), 507–518.
[30] Dorofeeva, A. A., Ponomarenko, E. A., Lukyanova, Y. Y., Fomina, E. A., & Buchatskiy, P. Y. (2021). High Precision Unmanned Agro Copters In Eco-Friendly Viticulture Systems. In *CEUR Workshop Proceedings* (pp. 299–306).

[31] Rodriguez-Fernandez, V., Menéndez, H. D., & Camacho, D. (2015, June). Design and development of a lightweight multi-UAV simulator. In *2015 IEEE 2nd International Conference on Cybernetics (CYBCONF)* (pp. 255–260). IEEE.

[32] Niu, Y., Zhang, L., Zhang, H., Han, W., & Peng, X. (2019). Estimating above-ground biomass of maize using features derived from UAV-based RGB imagery. *Remote Sensing*, *11*(11), 1261.

[33] Ashapure, A., Jung, J., Chang, A., Oh, S., Maeda, M., & Landivar, J. (2019). A comparative study of RGB and multispectral sensor-based cotton canopy cover modelling using multi-temporal UAS data. *Remote Sensing*, *11*(23), 2757.

[34] Horstrand, P., Guerra, R., Rodríguez, A., Díaz, M., López, S., & López, J. F. (2019). A UAV platform based on a hyperspectral sensor for image capturing and on-board processing. *IEEE Access*, *7*, 66919–66938.

[35] Gonzalez, L. F., Montes, G. A., Puig, E., Johnson, S., Mengersen, K., & Gaston, K. J. (2016). Unmanned aerial vehicles (UAVs) and artificial intelligence revolutionizing wildlife monitoring and conservation. *Sensors*, *16*(1), 97.

[36] Lin, Y. C., Cheng, Y. T., Zhou, T., Ravi, R., Hasheminasab, S. M., Flatt, J. E., ... & Habib, A. (2019). Evaluation of UAV LiDAR for mapping coastal environments. *Remote Sensing*, *11*(24), 2893.

[37] Kitpo, N., & Inoue, M. (2018, March). Early rice disease detection and position mapping system using drone and IoT architecture. In *2018 12th South East Asian Technical University Consortium (SEATUC)* (Vol. 1, pp. 1–5). IEEE.

[38] Tsouros, D. C., Bibi, S., & Sarigiannidis, P. G. (2019). A review on UAV-based applications for precision agriculture. *Information*, *10*(11), 349.

[39] Rasmussen, J., Nielsen, J., Streibig, J. C., Jensen, J. E., Pedersen, K. S., & Olsen, S. I. (2019). Pre-harvest weed mapping of Cirsium arvense in wheat and barley with off-the-shelf UAVs. *Precision Agriculture*, *20*(5), 983–999.

[40] Zhang, M., Zhou, J., Sudduth, K. A., & Kitchen, N. R. (2020). Estimation of maize yield and effects of variable-rate nitrogen application using UAV-based RGB imagery. *Biosystems Engineering*, *189*, 24–35.

[41] Marques, P., Pádua, L., Adão, T., Hruška, J., Peres, E., Sousa, A., & Sousa, J. J. (2019). UAV-based automatic detection and monitoring of chestnut trees. *Remote Sensing*, *11*(7), 855.

[42] Chartzoulakis, K., & Bertaki, M. (2015). Sustainable water management in agriculture under climate change. *Agriculture and Agricultural Science Procedia*, *4*, 88–98.

[43] Saccon, P. (2018). Water for agriculture, irrigation management. *Applied soil ecology*, *123*, 793–796.

[44] von Hebel, C., Reynaert, S., Pauly, K., Janssens, P., Piccard, I., Vanderborght, J., ... & Garré, S. (2021). Toward high-resolution agronomic soil information and management zones delineated by ground-based electromagnetic induction and aerial drone data. *Vadose Zone Journal*, e20099.

[45] Raj, R., Kar, S., Nandan, R., & Jagarlapudi, A. (2020). Precision agriculture and unmanned aerial Vehicles (UAVs). In *Unmanned Aerial Vehicle: Applications in Agriculture and Environment* (pp. 7–23). Springer, Cham.

[46] Gräf, M., Immitzer, M., Hietz, P., & Stangl, R. (2021). Water-stressed plants do not cool: Leaf surface temperature of living wall plants under drought stress. *Sustainability*, *13*(7), 3910.

[47] Kestur, R., Omkar, S. N., & Subhash, S. (2020). Unmanned aerial system technologies for pesticide spraying. In *Innovative pest management approaches for the 21st century* (pp. 47–60). Springer, Singapore

[48] Hussain, S., Masud Cheema, M. J., Arshad, M., Ahmad, A., Latif, M. A., Ashraf, S., & Ahmad, S. (2019). Spray uniformity testing of unmanned aerial spraying system for precise agro-chemical applications. *Pakistan Journal of Agricultural Sciences*, *56*(4), 897–903.

[49] Jorge, J., Vallbé, M., & Soler, J. A. (2019). Detection of irrigation inhomogeneities in an olive grove using the NDRE vegetation index obtained from UAV images. *European Journal of Remote Sensing*, *52*(1), 169–177.
[50] Sobayo, R., Wu, H. H., Ray, R., & Qian, L. (2018, April). Integration of convolutional neural network and thermal images into soil moisture estimation. In *2018 1st International Conference on Data Intelligence and Security (ICDIS)* (pp. 207–210). IEEE.
[51] Jung, J., Maeda, M., Chang, A., Landivar, J., Yeom, J., & McGinty, J. (2018). Unmanned aerial system assisted framework for the selection of high yielding cotton genotypes. *Computers and Electronics in Agriculture*, *152*, 74–81.
[52] Kellenberger, B., Marcos, D., & Tuia, D. (2018). Detecting mammals in UAV images: Best practices to address a substantially imbalanced dataset with deep learning. *Remote sensing of environment*, *216*, 139–153.

8 A Smart WeeDrone for Sustainable Agriculture

R. Maheswari, R. Ganesan, and Kanagaraj Venusamy

CONTENTS

8.1 INTRODUCTION

The major problem in present-day Indian agriculture is that the farmers are not able to look after their large hectares of land in an effective manner. In case of any disease outbreak in crops, the farmers are unaware of them and they become aware only when they spread and destroy large amounts of crops. The present role of IT in agriculture is guiding the farmers to take quality and effective decisions in order to increase crop yields with the guidance of experts sitting on the other side of the computer. However, in the agricultural field, these IT-based solutions could be applicable only to a certain level because, in some circumstances, the experts should analyze and identify the actual cause of the crops being affected by having on-site observation at the crop field. This will become hectic for both farmers and experts.

DOI: 10.1201/9781003252085-8

The same problem can be overcome when we bring the large field under real-time surveillance. The use of pesticides and insecticides is the major hindrance to sustainable agriculture since when these chemicals mix up with soil and water, it emits harmful effluents and affects the livestock and living.

8.1.1 Competition Between Crops and Weeds

In general, weeds are unwanted plants that grow in an agriculture field and compete for nutrients, moisture, light, and air. This results in a decrease in crop yield and crop growth due to nutritional deficiency that ultimately leads to disease outbreaks and puts overall productivity down. Table 8.1 shows the amount of loss of yield India is facing each year [1].

Reduction in yield by weeds accounts for 45%, whereas the insects 30%, diseases 20%, and other pests 5% [1]. Several methods are followed by farmers in order to control the weeds like cultural methods obtained less efficiently, manual weeding leads to high expense and it may be difficult in finding labor. Moreover, heat treatment ingests more energy, which leads to input costs. Soil solarization is a hydrothermal process that brings about thermal and other physical, chemical, and biological changes in the moist soil during and even after mulching. Modern weed control chemical methods are attaining popularity among farmers. The use of agrochemicals is negatively realized by consumers and supermarket chains. Applying heavy doses of pesticides initiates the problem of resistance development in weeds. Another problem is the regular usage of one pesticide can modify the weed community. So, an alternate, the better method could be devised to overcome the above-said methods. The smart WeeDrone is one such alternative method that is proposed in this system.

8.2 RELATED WORK

At present, drones are used for surveillance of a large crop area that cannot be manually controlled or overlooked. Some of the surveillance drones that are used in

TABLE 8.1
Yield Losses Due to Weeds

Crop Name	Loss of Yield (%)
Rice	9.1–51.4
Wheat	6.3–34.8
Maize	29.5–74.0
Millets	6.2–81.9
Groundnut	29.7–32.9
Sugarcane	14.1–71.7
Cotton	20.7–61.0
Carrot	70.2–78.0

present-day agriculture are Pix4DMapper Ag, Parrot drones, etc. There are different types of imaging facilities that can be used to take image of a large area. It includes a visual sensor with a high-resolution, low-distortion camera that can be used to produce images or video, which is useful for aerial mapping and imaging. Multispectral sensors use cameras that can measure colours in various resolution bands and eventually provide images in different wavelengths (e.g. near-infrared). It is useful for monitoring various health issues that arise in a plant and helps in identifying diseased plants and weeds effectively. A hyperspectral sensor is a sensor that can be used for plant health measurement, water quality assessment, and vegetable index calculations. Thermal infrared sensors are used to detect heat signatures and livestock detection.

Nowadays, drones are being used for surveillance, but they are not used for pest control to reduce the spread of diseases. For stopping the outbreak of diseases, the functionalities of drones must be enhanced with artificial intelligence and neural networks techniques. Advancements in science and technology have opened greater horizons to increase crop yield as well as crop destruction. The WeeDrone is proposed to be designed in such a way that it finds the weeds and infected plants through image processing and also removes them from the ground effectively.

Farms today are bursting with engineering marvels, the result of years of automation and other innovations designed to grow more food with less labour. The various weed control methods that exist in agriculture are a) preventive measures and b) curative or control measures. The curative measures have further been classified into i) mechanical methods, ii) cropping and competition methods, iii) biological methods, and iv) chemical methods.

8.3 WEEDRONE

The system is proposed to develop an ecofriendly, smart WeeDrone that can be used for large-scale mitigation of crops that are affected by viruses or other insects, and it will help in the destruction of weeds. It is also used to identify the weeds and infected plants that can be uprooted from the crop field using image processing techniques. Further, it helps to destroy the identified weeds and infected plants using the robotic arm of the drones that have minimal or no side effects when compared to the commonly used chemicals such as pesticides and insecticides. Thus, to make the process completely autonomous in image processing, the dynamic threshold values are identified by a machine learning algorithm to either cut the weeds or leave them. The shortest path planning algorithm will be implemented for the drone to fly a minimal distance in order to reach out to a weed-grown area for removing it, dispose of the waste at the ground station, and finally, come back to the field to find the next weed area.

This system is mainly designed to focus on diseased or dead plant removal, which will not be possible manually in the case of large fields since timely removal of diseased plants can prevent the spread of diseases. It incorporates a flexible robotic arm that may be used for placing or spreading some powder or manure in the fields. Based on previous and current trends of weed growth, data analysis can be done, which can predict the next pattern of weed growth. A drone can then be asked to monitor those weed-prone areas more often.

8.4 FEATURES OF WEEDRONE

8.4.1 Shortest Path Algorithm

The shortest flight path planning algorithms for drone flying will be implemented in order to ensure a reduced time and distance for the drone to go to a weed, remove it, put it to the ground station, and come back.

8.4.2 OpenDroneMap

OpenDroneMap (ODM) could be used. ODM is an open-source toolkit for photogrammetric processing of aerial imagery captured by an unmanned aircraft. Typical drones use simple point-and-shoot cameras, so the images from drones, while from a different perspective, are similar to any pictures taken from point-and-shoot cameras, i.e., non-metric imagery. OpenDroneMap turns those simple images into three-dimensional geographic data that can be used in combination with other geographic data sets. OpenDroneMap's processing engine includes robust methods for image matching, digital surface modeling, and image mosaicking [2].

8.4.3 Machine Learning Algorithm

The threshold value is set to either cut the weeds or leave them. This makes the process completely autonomous, which is expected. But maybe sometimes some detected weed need not be removed, or WeeDrone might detect the wrong plant as a weed, especially during training the image processing machine learning algorithm. In such cases, user feedback may be included after the image processing phase.

8.4.4 Removal of Diseased/Dead Plant

The system, along with weed removal, will also focus on diseased or dead plant removal, which will also not be possible manually in the case of large fields. Also, timely removal of diseased plants can prevent the spread of diseases.

8.4.5 Sowing and Spreading Manure

A flexible robotic arm may be used for sowing the seed across the field and also used for spreading fertilizer and manure in the fields.

8.4.6 Data Analysis

Based on previous and current trends of weed growth, data analysis can be done that can predict the next pattern of weed growth. The drone can then be asked to monitor those weed-prone areas more often.

8.5 DESIGN AND WORKING PRINCIPLE

The need for unmanned aerial systems (UASs) is expected to increase by $82 billion in a span of 10 years from 2015 to 2025, creating over 1,00,000 new jobs. The main area of having a commercial profit for the drone manufacturers is surveillance over large areas, especially agricultural farms. Precision agriculture helps farmers to measure the health and eminence of crops by imaging soil moisture, use of fertilizers, herbicides, and water management [3]. The imaging is done by different kinds of cameras suitable for particular applications. A multispectral camera [4,5], which works on higher bandwidth (in the infrared region), is used to capture the images of the field with finer details. Figure 8.1 shows the flow diagram of the proposed smart WeeDrone. The drone path is set using GPS coordinates [5] before it starts, by which it finds the path during its flight. Once it reaches the mentioned coordinates, the WeeDrone takes pictures of the field as a part of surveillance [4,6]. Those pictures are processed with the help of MATLAB software, and the weed-affected areas (coordinates) are displayed and the total value of weeds present in that area is calculated. The threshold value (T.V) for weed growth is set by the user. If the calculated value is lower than the threshold value, then the drone just mails all the data (images and calculated value) to the user. If the calculated value is higher than the threshold value, the drone gets the coordinates of the weed-affected area from the MATLAB software. As a result, the WeeDrone reaches the targeted area, activating the robotic arm to pluck out the weeds from the field. Finally, the drone sends an email about all the completed work to the user.

8.6 OUTPUTS AND SIMULATION

The complete functionalities of the WeeDrone are divided into three modules as autonomous flight, image processing of the field, and design of robotic arm.

8.6.1 Autonomous Flight

The two major components of the autonomous flight module are drone flight and automatic flight subsystem. It includes automatic flight software, a front-facing camera for imaging, GPS (global positioning system) transmitter, an inertia measuring unit (IMU). The IMU consists of three accelerometers, three gyroscopes, and three magnetometers. A barometric sensor is used to maintain the altitude of flight along with a timing micro-controller and a Beagle Bone Black (BBB) micro-controller. BBB is a Linux-based multitasking system [3,6] where the autopilot software will be loaded. The need for precise flight control with respect to time is needed and is achieved by using the Atmel Atmega timing microcontroller.

8.6.1.1 Drone Flight

The drone that is being designed is supposed to be fully autonomous. The user can mark the surveillance area through GPS coordinates. As soon as the drone is launched, it finds its location using GPS and makes an autonomous flight [6,7] over the marked area and takes pictures using the multispectral camera. The images

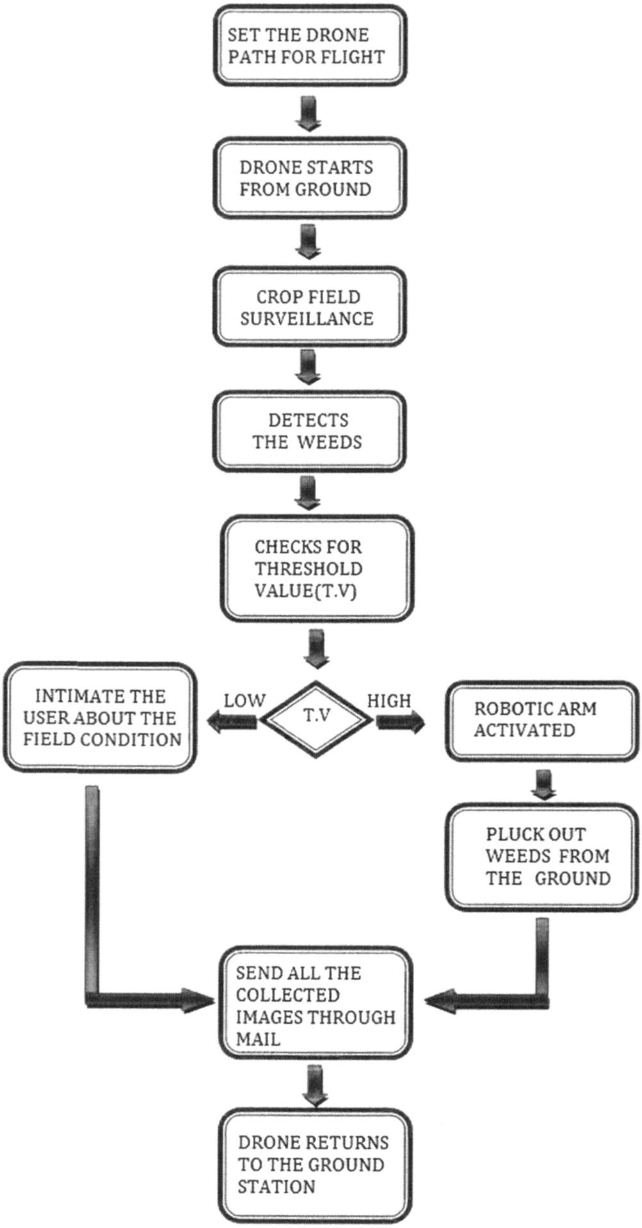

FIGURE 8.1 Flow diagram of WeeDrone.

taken by the drone are merged into a single large image that can be mailed to the user and ask their preferences for further processing. The drone has been designed in such a way that it can detect the obstacles such as towers that hinder their navigation path.

8.6.1.2 Automatic Flight Subsystem

This system is embedded on the BBB, a Linux-based single-board computer [3,6]. The GPS transmitter sends the coordinates to the BBB microcontroller and establishes the navigational route for the UAV. In case of emergency, the control will be given to the user in order to operate the drone by using a R/C (remote control) transmitter [4,7]. In turn, the transmitter sends a signal to the BBB microcontroller to stop; as a result, the drone comes under the control of the user. The front-facing camera is used to avoid obstacles in the navigation path using artificial intelligence techniques like pattern recognition, learning from experience and optical flow [8]. The timing microcontroller collects the information from IMU and communicates it to BBB. The automatic flight software [6], which has been loaded on the Beagle microcontroller, will adjust the navigation path of the drone with respect to the inputs received from other microcontrollers in order to achieve a perfect flight. Figure 8.2 represents the block diagram of the Automatic Flight Subsystem.

8.6.2 Image Processing

The multispectral camera (MS4100) is used for imaging over the crop fields that provide images with which area-wide pest management and weed control can be done effectively. MS4100 [4–6] which has image sensors with a charged couple device supporting three types of images that include color infrared (CIR); red green blue (RGB); and a combination of both either in the blue or red range. Figure 8.3 gives the acquired image from the MS4100 camera.

The obtained image undergoes various functionalities like field operations, data processing, image analysis, and feature extraction. The infrared image that is

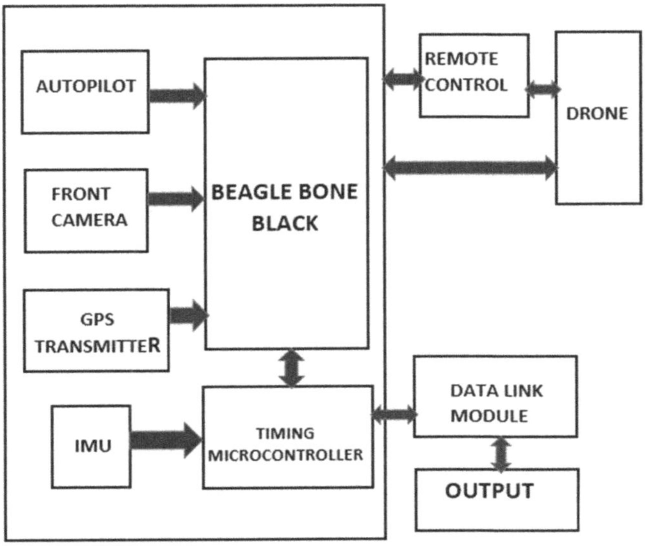

FIGURE 8.2 Block diagram of automatic flight subsystem.

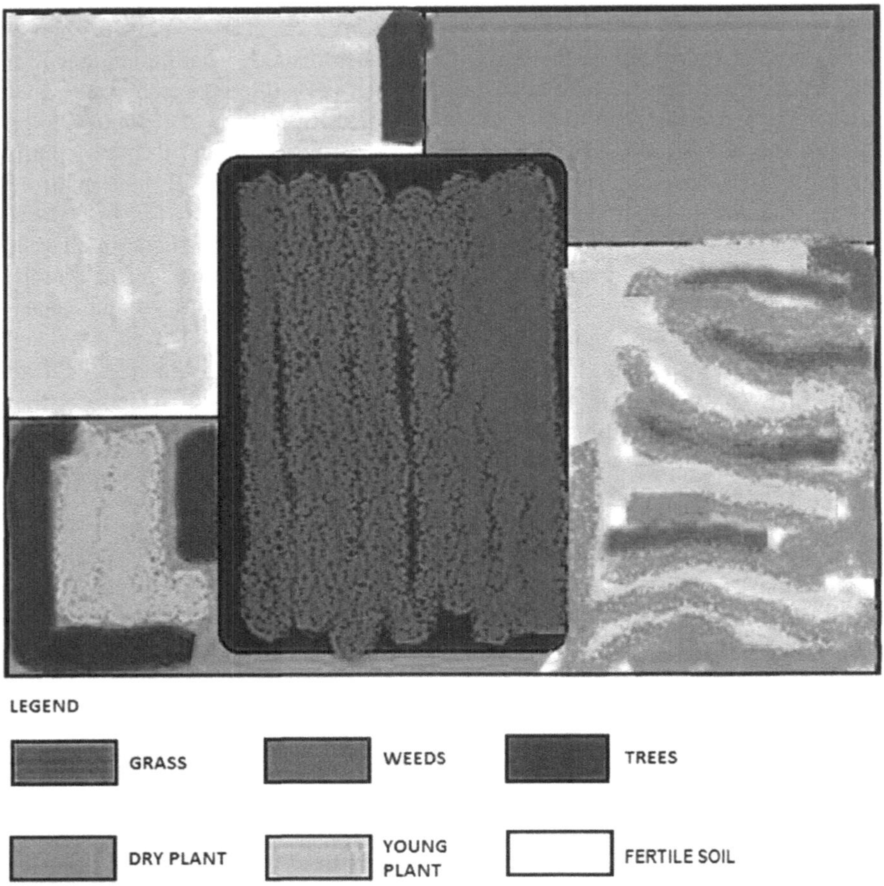

FIGURE 8.3 Image obtained from MS4100 camera.

acquired from the camera is transformed to a bi-level (black and white) image or simply a binary image [9,10] with the help of MATLAB software. A threshold value is set to differentiate between healthy crops and weeds. The infrared camera captures the image based on the temperature that is reflected back from the field. The areas where the temperature is higher than the threshold are changed to white pixels representing the weeds or unhealthy crops that have to be removed from the agriculture field. The areas with lower temperatures are converted to black pixels representing the healthy crops. Figure 8.4 shows the image obtained from MATLAB software.

The white region represents the weed-affected area. The coordinates (waypoints) of the white region are derived from the MATLAB and given as input to the BBB microcontroller so that the drone navigates to that particular location and removes the weeds using a robotic arm. Image processing for field operations, data processing, image analysis, and feature extraction are deeply evaluated [11].

FIGURE 8.4 Image obtained from MATLAB software.

8.6.3 Robotic Arm Design

The robotic arm is designed in such a way that it can follow the nonlinear path [12,13] using waypoint GPS and reach the target area where the amount of weeds is greater than the threshold value. Now the work of the robotic arm is to pluck out or remove the weeds from the field and place it in the nearby waste ground. Figure 8.5 shows the robotic arm designed for WeeDrone.

8.6.3.1 Statistical Analysis of WeeDrone Model

A nonlinear control method [13,14] is designed in such a way to achieve an optimized performance by the drone. In general, the aerial system can be broadly classified into coupled and decoupled models. The difference is that in a decoupled model, the external forces or disturbances have to be compensated by the drone itself. The optimal control algorithm [15] is used for deriving the reference trajectories so that the weeds can be taken from the field and can be dropped in the nearby wasteland. The drone is designed by connecting n+1 rigid body with v_i, representing its linear velocity and w_i, representing the angular velocity of the drone [12,16]. The inertial tensor is denoted as

$$I_i = (J_i 0 \;\; 0 \;\; m_i I_n) \tag{8.1}$$

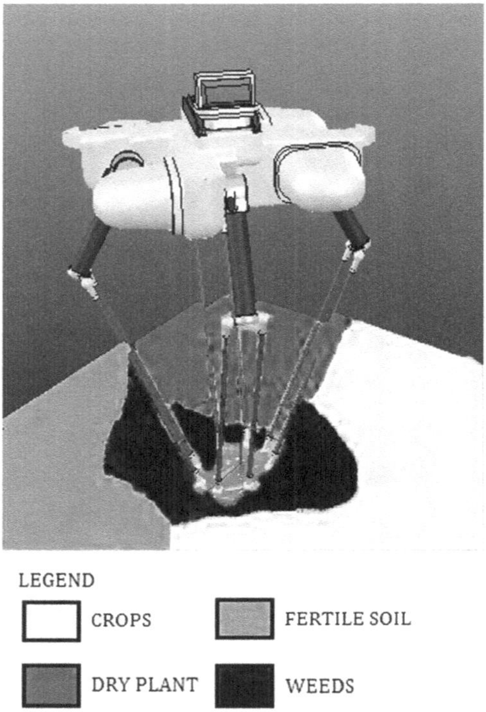

FIGURE 8.5 Robotic arm embedded in WeeDrone.

J_i = rotational inertia tensor, m_i = mass of the drone, and I_n = identity matrix in Eqn. 8.1.

$$\hat{v} = (\hat{w} \;\; v \;\; 0_n \;\; 0) \tag{8.2}$$

$\hat{v}$ represent the base linear velocity over a given frame g in Eqn. 8.2.

$$\hat{w} = (0 - w_3 \;\; w_3 \;\; w_3 \;\; 0 \;\; - w_1 - w_2 \;\; w_1 0) \tag{8.3}$$

The corresponding base angular velocity $\hat{w}$ over a given frame g is given in Eqn. 8.3.

Continuous equations of motion and spatial operator theory can be used to conclude that the mass depends only on the shape variables and not on the joint variables [12]. The remove and throwaway motion can be achieved using the model predictive control to get the future trajectories between the interval (t_0, t_f), where t_0 is the current time and t_f is future trajectory time [17,18]. The optimal predictive control can be implemented using Gauss-Newton Shooting method and Stage-wise Newton method [13]. The Gauss-Newton Shooting method can be used to achieve the standard regularization and line-search techniques using retract and lift operators [12,19]. Using these methods, the reference trajectory for the entire WeeDrone

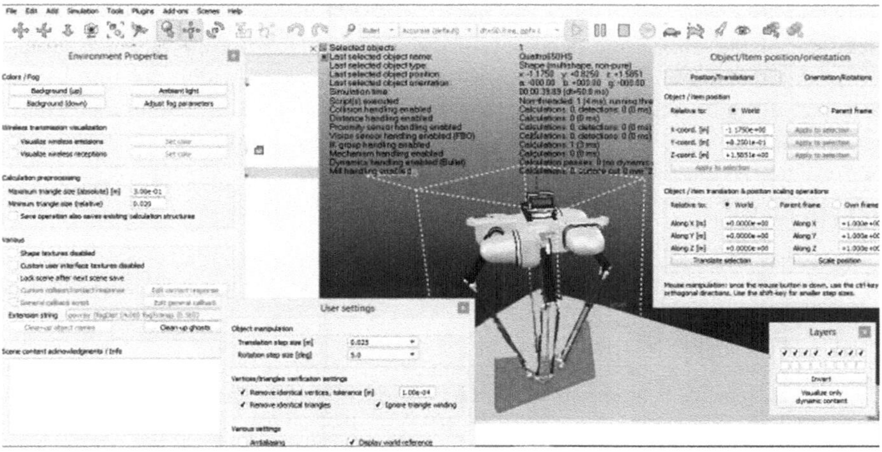

FIGURE 8.6 Simulation of WeeDrone.

is determined that includes both UAV and robotic arm. The determined trajectory contains all the states of the drone, which includes the position, orientation, linear, and angular velocities. It also contains all the states of the robotic arm which includes the joint positions and their velocities. The optimal control approach helps the arm to move faster by maintaining accuracy. As a result of this approach, the trajectories of both drone and arm are executed concurrently and the weeds are plucked out quickly [20,21]. Figure 8.6 represents the simulation of WeeDrone that takes its path to the weed-affected area and plucks out the weeds successfully.

Figure 8.6 represents the simulation of WeeDrone that takes its path to the weed-affected area and plucks out the weeds successfully.

8.7 CONCLUSION

Thus, this proposed system ensures the development of an eco-friendly intelligent drone named WeeDrone for large-scale mitigation of crops that are highly affected by viruses or other insects. This design helps in the destruction of weeds in an automated way which forms an operational method to destroy weeds with an automatic robotic arm. The system also enables evaluating image processing procedures to separate weed and crop plants using multispectral imagery. Using the shortest flight algorithm, the system identifies the weeds appropriately and activates the robotic arm for Weedrone removal. This system not only removes the weed but also removes diseased or dead plants, which is not possible manually in the case of large fields and moreover, timely removal of diseased plants can prevent the spread of diseases. A flexible robotic arm can also be used for sowing the seed across the field and also for spreading manure in the fields. This system guarantees minimal side effects or zero side effects since it prevents or reduces the use of chemical pesticides, confirming sustainable agriculture. WeeDrone may be extended to perform field surveillance which may check for areas of waterlogging and other problematic areas in agriculture. As the accuracy of the weed map is based on the precision of the image and accuracy of the analysis, appropriate advanced image processing technique. Since the size of the crop field varies widely, battery backup becomes a major challenge. Thus, to promote sustainable agriculture in greater ways, WeeDrone shall be driven with a solar panel for power supply.

REFERENCES

[1] Rao, V. S., and Mohan Primlani, *Principles of Weed Science*. 2nd edition, published by for Oxford and IBH Publishing Co. Pvt. Ltd., New Delhi, 2000, Vol. 2, pp. 1641–1647.

[2] http://opendronemap.org/

[3] Huang, Y. Lan, and W. C. Hoffmann, Use of Airborne Multi-Spectral Imagery in Pest Management Systems published in Agricultural Engineering International: the CIGR Ejournal. *Manuscript IT 07 010*. Vol. 10. February 2008, pp. 161–200.

[4] Yang, Chenghai, John K. Westbrook, Charles P.-C. Suh, Daniel E. Martin, W. Clint Hoffmann, Yubin Lan, Bradley K. Fritz, and John A. Goolsby, An Airborne Multispectral Imaging System Based on Two Consumer-Grade Cameras for Agricultural Remote Sensing, *Remote Sensing*, Vol. 6, 2014, pp. 5257–5278.

[5] Burai, P., J. Tamas, C. S. Lenart, and I. Pechmann, Usage Of Different Spectral Bands In Agricultural Environmental Protection in Journal of Agriculture, 2013, pp. 1–4.

[6] Sinsley, Gregory L., Lyle N. Long, Albert F. Niessner, and Joseph F. Horn, Intelligent Systems Software for Unmanned Air Vehicles at 46th AIAA Aerospace Sciences Meeting, Reno, NV, AIAA, 2008, pp. 71–80.

[7] Lee, Dr. Richard, and Dr. Roberto Desimone, Flexible Uav Mission Management Using Emerging Technologies in 2002 Command and Control Research and Technology Symposium, US Naval Postgraduate School, Monterey, California, 11-13th, June 2002.

[8] Chithapuram, Chethan, Yogananda V. Jeppu, and Ch. Aswani Kumar, Artificial Intelligence Guidance for Unmanned Aerial Vehicles in Three Dimensional Space, International Conference on Contemporary Computing and Informatics, pp. 1256–1261, 2014.

[9] Zhang, Huihui, Yubin Lan, Ronald Lacey, W. C. Hoffmann, and Yanbo Huang, Analysis of Vegetation Indices Derived from Aerial Multispectral and Ground Hyperspectral Data, *International Journal of Agricultural and Biological Engineering*, Vol. 01, 2009, pp. 2–4.

[10] Lippiello, Vincenzo, and Fabio Ruggiero, Cartesian Impedance Control of a UAV with a Robotic Arm at 10th IFAC Symposium on Robot Control International Federation of Automatic Control September pp. 5–7, 2012.

[11] Landge, Pramod S., Sushil A. Patil, Dhanashree S. Khot, Omkar D. Otari, and Utkarsha G. Malavkar, Automatic Detection and Classification of Plant Disease through Image Processing, *International Journal of Advanced Research in Computer Science and Software Engineering*, Vol. 3, Issue 7, 2013, pp. 798–801.

[12] Garimella, Gowtham, and Marin Kobilarov, Towards Model-predictive Control for Aerial Pick-and-Place in IEEE International Conference on Robotics and Automation (ICRA), pp. 4692–4697, 2015.

[13] Bertsekas, D. P. *Nonlinear Programming*, 2nd ed. Athena Scientific, Belmont, MA, 2003.

[14] Jain, Abhinandan. *Robot and Multibody Dynamics: Analysis and Algorithms*. Springer, Pasadena, USA, 2011.

[15] Murray, Richard M., Zexiang Li, and S. Shankar Sastry. *A Mathematical Introduction to Robotic Manipulation*. CRC, Berkeley, 1994.

[16] Ruggiero, F., M. A. Trujillo, R. Cano, H. Ascorbe, A. Viguria, C. Perez, V. Lippiello, A. Ollero, and B. Siciliano, A multilayer control for multirotor UAVs equipped with a servo robot arm in IEEE International Conference on Robotics and Automation, pp. 4014–4020, 2015.

[17] Arivazhagan, S., R. Newlin Shebiah, S. Ananthi, and S. Vishnu Varthini, Detection of unhealthy region of plant leaves and classification of plant leaf diseases using texture features by, *CIGR Journal*, Vol. 15, Issue 1, 2013, pp. 211–217.

[18] Jun, W., and S. Wang. Image Thresholding Using Weighted Parzen Window Estimation. *Journal of Applied Sciences*, Vol. 5, 2008, pp. 772–779.

[19] Patil, Girish, and Devendra Chaudhari, SIFT Based Approach: Object Recognition and Localization for Pick-and-Place System in *International Journal of Advanced Research in Computer Science and Software Engineering*, Vol. 3, Issue 3, 2013, pp. 196–201.

[20] Avellar, Gustavo S. C., Guilherme A. S. Pereira*, Luciano C. A. Pimenta, and Paulo Iscold, Multi-UAV Routing for Area Coverage and Remote Sensing with Minimum Time, sensors ISSN 1424-8220, Vol. 15, Issue 11, 2015, 27783–27803.

[21] Almalki, Faris A., Ben Othman Soufiene*, Saeed H. Alsamhi, and Hedi SakliA, Low-Cost Platform for Environmental Smart Farming Monitoring System Based on IoT and UAVs. *Sustainability*, Vol. 13, Issue 11, 2021, pp. 5908–5912.

9 Internet of Agro Drones for Precision Agriculture

C. Muralidharan, Mohamed Sirajudeen Yoosuf, Y. Rajkumar, and D.D. Shivaprasad

CONTENTS

DOI: 10.1201/9781003252085-9

9.1 INTRODUCTION

Agriculture is one of the important fields that everyone directly or indirectly depends upon. With the increasing supply throughout the world and the lowest commodity prices, the consumption and production of agricultural products tends to increase tremendously and that becomes a crossroad in the modern agricultural industry [1–3]. Not only is there a demand for agricultural commodities, there is a high level of demand for the farmers and well-practiced agronomists for managing the agricultural resources. Apart from these issues, another factor that has a direct impact over the agricultural field is climate change, which creates complexity in protecting the supply chain security of the agricultural industry [4]. Also, other environmental conditions that evolve rapidly further increase the complexity. Hence, priority has to be given for optimizing the sustainability so as to reduce the impact as well as public and planet well-being. Because the improved sustainability support the agricultural practitioners with additional benefits in their economics and also to manage the resources efficiently. Intergovernmental Panel on Climate Change provided a report, which states that managing the land is one of the important processes for overcoming the drastic climate change so that the land degradation can be reduced. The report also states that the policy for managing the land should be framed in a manner that it should be cost effective and the benefits have to be exercised for a long time. Every practitioner should keep in mind that the farms not only have to withstand the drastic climate change but also to withstand the economic loss [5]. There is a movement named "farm to fork" that concentrates more on traceability of every agricultural products as the consumers are interested in knowing the details about the products, such as who produced it and how they are grown. From all these constraints, it is clear that the practice of agriculture has to be refined in a precise manner. Hence, there is a need for us to concentrate more on precision agriculture.

Precision agriculture is a type that enriches the management of farming through regular observation of the crops and measures the variability, thereby responding to it in a proper time [6]. The main aim of precision agriculture is providing a decision support system through which the farm will be managed by optimizing the returns and thereby preserving the resources. The usage of tools such as global positioning systems and the geographic information system allows us to monitor and map all the parameters that are related to the crop, which generates useful data that is used for intensifying the cultivation methods. These tools support the farmers in identifying the level of fertilizers at the needed time and support them in identifying the disease on the crop before it spreads over the entire land. These collected data support the farmers in identifying the crop that is suited well for the land based on the seasonal changes. This also supports them in making decisions based on the environmental as well as economic factors. This precision agriculture is enlightened due to the emergence of drones. The drones are electronic-based machines that can be controlled remotely for automating the agriculture processes. They reduce the manual work and speed up the processes [7,8]. These drones are installed with different sensors like cameras, global positioning systems, etc. through which the crops are monitored from a remote place. The usage of drones tends to increase in the agricultural industry, which supports the farmers, agronomists, and agricultural engineers in identifying the insights of crop cultivation through proper data analytics and for managing the agricultural operations. For example, crop monitoring is one of the processes that is made easier by analyzing the data that are collected by the drones for making the accurate plan or needed improvements such as suggesting the fertilizer in a timely fashion, indication of disease-prone plants, and so on. Further, all the agricultural products can be traced completely right from the farm to the end user through the utilization of GPS, which is a more efficient model when compared to the traditional manual information acquaintance model.

The manual monitoring model has lots of difficulties, such as handling large farms, analyzing farms at slope areas, and so on. Hence, these drones are most suitable for handling these types of processes as they can travel any place based on the way we handle it. They can work at any type of land at any place. Not only with the use of drones, the monitoring and analysis can also be done with other IoT equipment that has to be installed at the agricultural lands. The drawbacks with this equipment are they can be disturbed by any animals, large areas of land require more equipment, power resource for the equipment, etc. But in the case of drones, the single drone can monitor the large area from the remote place through which we can overcome the drawbacks. More crucially, the data collected by the drones is high resolution in nature, which can be used to access the information about the level of fertility of the crops that is used for managing the irrigation system and further reducing waste.

Not restricted with the monitoring process is the unmanned aerial vehicle (UAV), which supports in improving the sustainable agriculture. It's found that the worth of drones in agriculture is USD$32 billion, which shows that people have recognized the use of UAVs [9]. Apart from the support of making decisions with regards to agricultural works, it is considered the safest mode of mapping agricultural areas when compared to other models because it can handle uneven areas, making it free from hazardous activities to the operators. While compared to manned aircrafts and the satellite-based monitoring, UAVs are cost effective as well as accurate in capturing the data from the environment. Also, another important feature that exists with the UAVs is that even in bad weather conditions it collects the data in an accurate manner when compared to the manned aircrafts. The pictures of the different parts of the plants are captured clearly, which may support the farmers in identifying the diseases earlier [10]. Hence, there is a need for the farmers to use the drones for employing precision agriculture. Once they decide to use it, a few factors have to be considered i.e., choosing the type of drones to use for the agricultural processes. There are two different types of drones: rotary and fixed wing. These types have several advantages and disadvantages of using it for agricultural processes, which is further discussed in section 9.3. Thus, UAVs are the key consideration for farmers practicing precision agriculture, which may support the farmers in improving the farming and increasing the revenue by it. This chapter discuss the need for precision agriculture and the use of drones with their applications.

Section 9.2 discusses the need and importance of precision agriculture, section 9.3 discusses about the types of precision agriculture, section 9.4 discusses the drones' contribution in precision agriculture, section 9.5 discusses the use cases of drones in agriculture, section 9.6 discusses the future of agricultural drones, and section 9.7 concludes the chapter.

9.2 PRECISION AGRICULTURE

The interest in agriculture is increased throughout the world nowadays. Not depending on the agricultural land, farming has been practiced even on the tops of houses. It's all because of the need for people to get pure products. To make an agricultural product in a pure form and to increase the productivity, regular monitoring and a timely process are needed [11]. This can be done with the emergence of precision agriculture. The following sections discuss precision agriculture.

9.2.1 What Is Precision Agriculture?

Precision agriculture is the form of practicing farming in a precise manner that supports the farmers in making decisions for managing the crops and to improve the crop yields with the use of multiple sensors [12]. Throughout the world, farmers are showing their interest towards precision agriculture as it supports them in increasing agricultural products, effective management of crops through proper fertilization, reduces labor needs, irrigation management, and so on. This type of agricultural practice uses a large amount of agricultural data that is collected from the lands. These data are processed for understanding the growth of plants. It is one of the advanced models that acquires the data through direct interaction with the plants and is analyzed for improving the yield of resources. The information is collected through the use of UAVs, and it is possible to collect multiple data at a time, such as information about irrigation, growth levels of plants, disease attacks, and so on. With high spatial resolution, they monitor the health of the crops and collect information that is further implemented for the same type of plants when they are planted for the next time. Ultimately, the main aim of precision agriculture is to increase the productivity of the agricultural products as well as the revenue of farmers.

9.2.2 Importance of Precision Agriculture

Traditionally, agricultural production is based on climate conditions and farmers do practice agriculture based on the seasons. Many improvements have been immersed in the field of agriculture that any crop can be cultivated at any time. But this will reflect in the crop yields that may produce the products with less taste and so on due to the drastic change in the climate conditions. Hence, the normal flow of farming collapses and proper analysis is needed for identifying the proper crops that can be cultivated at proper times [13]. This analysis needed proper data that has to be acquired from the crops that are cultivated in a real-time fashion. Along with this, the climate conditions and their changes have to be monitored. Nowadays, the production of agricultural products completely relies on monitoring the status of the crops right from the health, irrigation, soil condition, disease attacks, and so on, which seems to be a major challenge for farmers. This can be done with the implementation of precision agriculture, which uses several sensors along with UAVs. The multiple sensors installed in the UAV collect the data from different sources parallelly. The collected data are then stored and analyzed for extracting the useful information about the crops. This obviously increases the production of crops and thereby increases the revenue for farmers. Another major benefit in using the UAVs is that they can be operated from a remote place and are able to monitor even in slope regions, which is a major problem that most farmers face in hilly stations.

9.2.3 Advantages of Precision Agriculture

There are several advantages in implementing precision agriculture. A few are mentioned below:

- The real-time data can be collected based on the variables that occur in the agricultural land.
- Efficient in managing the irrigation system, thereby reducing the total water consumption.
- Reduces the number of laborers needed for the work, which cuts costs.
- Provides regular monitoring that increases the quality of the product.
- It gives alarms about climate changes such as rain, dryness, etc.
- Avoids manual monitoring of the agricultural products.
- Reduces the environmental impact, thereby regulating the use of agrochemicals.

9.3 DRONES' PART IN PRECISION AGRICULTURE

The unmanned aerial vehicles have created a revolution in agricultural industry. Drones are one form of unmanned aerial vehicle that are used for agricultural purposes in the view of optimizing the crop yield and also to monitor the growth as well as the production of the crops [14]. There are many purposes in using drones in agricultural processes, such as monitoring the crops, fertilizing the crops, monitoring the soil, giving alarms about the climate changes, regulating irrigation, and so on. This section discuss the drones' part in agriculture.

9.3.1 Need and Use of Drones in Agriculture

Monitoring is one of the important processes that needs to be done very frequently in the case of agriculture. If a farmer fails to monitor the different stages of crops, then the expected yield cannot be acquired. Also, the manual process of monitoring the crops is not easy, as different crops have different growing patterns and even the number of days needed for their growth varies depending on the crop type [2,12]. Hence, for resolving this type of issue, agricultural drones can be used for auto-monitoring of crops in a parallel manner as they can be installed with different sensors. They are not limited to the monitoring process because they can also be used for fertilizing the crops. Different crops will be different sizes, which increases the complexity in fertilizing the crops manually by the farmers, but if the drones are used, the fertilizers can be sprayed in the crops easily by the above surface and literally reduces the time needed for the entire process.

Drones can be used for capturing all the processes that happen on the agricultural lands such as the soil level, type of crops grown, density of crops, disease attacks on the plants, different stages of plant growth, etc. Manually analyzing all these things is not easy and it needs more laborers and time, which obviously increases the total costs [4,13]. But when the drones are used, they capture the data from higher altitudes at different positions, which further can be collected in different locations for analyzing the data. This analyzed data can be utilized for improving the cultivation of the crops. Another important feature in using the drones is that monitoring or fertilizing the crops that are cultivated in slope regions can be possible, which is a tedious process manually.

9.3.2 Best Agricultural Drone Practices in Precision Agriculture

There are several best practices in using the drones in agriculture. They are as follows:

- Irrigation management

  The drones captures the data about the soil and the humidity is checked frequently. Based on the humidity level, the irrigation system is invoked, thereby reducing the water consumption.
- Surveillance and crop health monitoring

  The crops will be monitored by capturing several images at different positions. Images that are captured at all the regions are compared with each other for checking the growth level of crops.
- Disease assessment

  The images captured at different positions are checked for disease or pest attacks for alarming the farmers about the spread of diseases. This further calculates the level of spread on the land.
- Soil analysis

  The soil conditions can be tracked through the survey made by the drones with the help of multispectral sensors. The chemical content in the soil can be calculated and the need for fertilizer can be suggested.

- Controlled fertilization
  The drones are now used as sprayers through which the fertilizers needed for the crops can be sprayed by controlling them from remote places. This reduces and regulates the use of fertilizers.
- Livestock tracking

The drones can be used for tracking livestock easily from remote places. It is too hard for the farmers in tracking the livestock when they live near the forest, which can be done through the drones.

9.4 TYPES OF PRECISION AGRICULTURE

There are several types of precision agriculture. The following sections discuss the types in detail.

9.4.1 Precision Seeding

Precision seeding is described because of the location of favored numbers of seeds at a specific depth and spacing. Precision seeding has many boosts for the vegetable grower over traditional dribble (Planet Jr.) or multi-seed drop-plate seeding strategy (nearly all corn planters). On the other hand, the seeding accuracy isn't always a replacement of appropriate land cultivating, irrigation, and distinct crop manage practices are essential to acquire an amazing stand of a vegetable crop. The precision seeding types are shown in Figure 9.1

There are various types of seeders.

9.4.1.1 Belt Type

A belt, which is has holes, will help with seed size and also to maintain the interval between the seed holes. Examples: Tomato, watermelon, etc.

9.4.1.2 Plate Type

Seeds drop whit in notch in the saucer and later it is moved to the drop portion . The largest spacing is finished with the aid of gearing the proportion of the propel of the saucer. Examples: Lettuce, snap beans, etc.

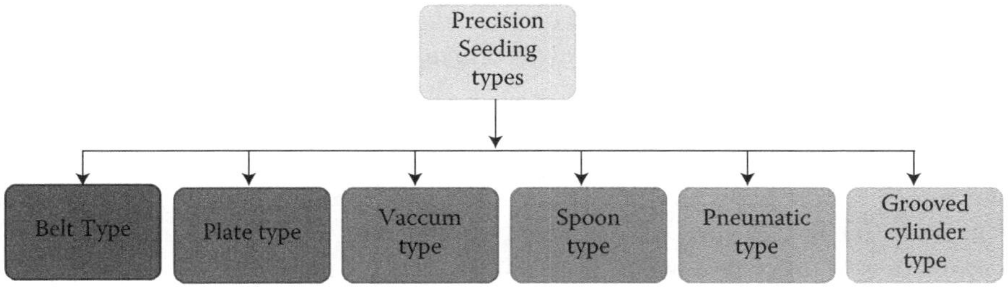

FIGURE 9.1 Precision seeding types.

9.4.1.3 Vacuum Type

A seed is drawn towards a hole in a vertical plate and is agitated to put off extra seed through an aggregate of gears and hole quantity in keeping with plate, numerous spacings are accomplished. Example: Watermelon seeds, snap beans, etc. It can handle large and small seeds also.

9.4.1.4 Spoon Type

Here, the spoons are available according to the seed sizes. A seed is moved up out of a reservoir by using small spoons (sized for the seed) after which it is carried to a drop shoot in which the spoon turns and drops the seed.

9.4.1.5 Pneumatic Type

A seed is held in location in opposition to a drum until the air stress is broken, and then it drops in pipes and is windblown across the soil.

9.4.1.6 Grooved Cylinder Type

The cylinder pops slowly and when it reaches the lowest of the case, the seeds spill out of a diagonal slot. By mixture of forward pace and turning rate, the seed is located at preferred increments. This planter may be used with seeds no larger than a pepper and it works excellently with a coated seed.

9.4.2 Precision Soil Preparation

Historically, the goals of soil sampling had been to identify the normal nutrient repute of a subject and to offer a few measures of nutrient variability in a discipline. Sampling can be done by two ways like I) grid sampling and II) direct sampling.

9.4.2.1 Grid Sampling

An earlier system consisting of limited domestic animals, heavy manure utility, or competitive leveling for irrigation has drastically altered the soil nutrient stage. Small fields with specific cropping histories were merged into one. This became a base map of natural soil.

9.4.2.2 Direct Sampling

The yield maps, remotely sensed pictures or various assets of spatial data, are used and display consistency from one layer to other. When we practice farming, the field that you are feeling might offer a path on which to delineate control zones. There may be restricted or no history of farm animals or manure having an impact on the sector.

9.4.3 Precision Crop Management

Integration of aerial imagery and more than one sensor technology is needed to deliver a 360-degree view of the plant surroundings close to actual time. 24/7 transmission of facts through mesh/mobile network capacity and the supply and evaluation of high-quantity sensor information of advanced crop productiveness (yield) can be used as research information. This further be used for predicting early abnormality and corrective moves functionality naming of superior (and sub-premiere) crop situations with actionable perception . The middleware will discover the capability of the IoT era to enhance crop control through various manufacturing (yield), minimize operational expenses, plus smarter packages of chemical compounds and fertilizers. The middleware will enhance crop yield via the statistical of actual-time records from a selection of environmental sensors and different resources of reality placed in business crop fields or at some stage in the corporation.

9.4.4 Precision Fertilizing

In most of the cases, the farmer will know the details about their own farm, such as which part is more fertile and which part is not. Because they have a close association with that farmland, once we are aware of this information, we seed them accordingly. With the information that was captured from a satellite image for the field and process, the images generate an optimum solution to improve fertilizer rate.

Optimum fertilizer practices majorly depend on the factors such as:

9.4.4.1 Target Metrics

There may be either straight or in a roundabout way associated with a few variables that can be used to estimate plant nutrient requirement.

9.4.4.2 Bounded Field Metrics

The result that was extracted from the regular practices are compared with the targeted results in the form of matrices.

9.4.4.3 Contemplate with the Temporal Situations

It will help to find the crop growth, yield rate, consumption of water, etc.

9.4.4.4 Study about the Responses from the Crop

After applying the strict metrics and targets, we can study the response of the crop and make the changes accordingly.

9.4.5 Precision Harvesting

The farmers, before they are going to yield or if they were initiating for the plantation, will check with the information brochure, in that the farmer will get all the required information regarding their crop, because this brochure was prepared by the organization, which follows precision farming methodologies. By following these practices, the farmer will benefit with good profits by applying less water, less fertilizer, etc.

9.5 USE CASES OF AGRICULTURAL DRONES

There are multiple case studies that prove the use of agricultural drones to a greater level, where the advantages of a farmer are beyond our imagination and credits. Presented here are few of the case studies where the agricultural drones are used.

9.5.1 Case Study: Reducing Herbicide Use in Brazil

With the help of drones, the use of chemicals for killing herbicides has been reduced to a considerable extent of 52% in Brazil, where the soyabeans are cultivated. Within a span of one and half hours, the drone flight mapped a maximum of 500 hectares (nearly 1,200 acres), which is of 100% accurate and high quality. These images are analyzed and it captured exactly how far the plants are affected by weed infestation. This helps to process the image and thereby reduce the use of herbicide where mappings can be utilized for further study in 2018 and 2019.

9.5.2 Drones – VTOL and VTOL Hybrid Drones

Flying eye drones are the most occupied and used types that are fixed with sensors, cameras, batteries, and radio controls. They are actually used for spraying the crops, by taking images and videos pertaining to the far-flung areas, thereby making the customer safe.

9.5.3 Japan's Suitable Point of Reference

Being one of the largest agrochemical manufacturers in the world, Japan has used remote-controlled helicopters. This produces a quantitative nature of the data on the fields for the past 30 years. The companies in Japan started investing in drone technology with Nileworks based in Tokyo, which accrues funding from the United States. This has considerably reduced the workload on farmers.

9.5.4 Using Drones to Spot Disease and Weed Infestations in Sugar Beets

An agronomist and drone operator teamed up to find better ways of detecting plant stress in sugar beets and spotting disease and weed infestations with drones. A sugar beet grower, David, had heard about the potential of drone technology from a friend who had been using drones to provide counts of plant and tree populations. A few weeks before harvest, David, a farmer and agronomist, approached Agremo for advice on the use of drones with sugar beets, specifically about how they could be used to detect plant stress. Drones can provide plant counts, data on the location of certain weeds and diseases, or identify irrigation problems by identifying areas of water stress. Thus, they represent an exciting opportunity for farmers to improve crop management and allow a more targeted use of inputs. Using drone imagery combined with Agremo plant stress analysis, David was able to produce a map indicating areas of weed and disease infestation within the crop. David could consider spraying only the affected areas of the field or using VRT (variable-rate technology).

9.6 FUTURE OF AGRICULTURE DRONES

In any industry, innovation plays a key factor in tackling the challenges in the present as well as the future. According to a survey, it has been estimated that nearly \$6.7 billion has been invested during the past five years in the agricultural industry. Several different kinds of innovations, such as vertical farming, indoor vertical farming, automation, robotics, greenhouse farming, precision agriculture, etc. have revolutionized the farming industry like never before. Across these arenas, technologies like artificial intelligence, machine learning, drones, Internet of Drone Things, smart drones, 5G, and blockchain have enabled the farming practices to be imagined in the near future. Presented here are a few practices that have been under the lab, which are going to be before our eyes in the near future.

9.6.1 Application of 3D Printing Technology to Food

3D printing a most prominent technology in the industry of manufacturing and is extending its reach to food production. This is called as additive manufacturing, where the exact material can be created as objects. Several researchers have been on the belief that the usage of hydrocolloids (where the substances that forms gels with water) can be utilized as a base for the production of foods by using reusable algae, duckweed, and grass. An applied science research organization based on Netherlands came up with a new innovative procedure to develop microalgae, a natural source of protein, carbohydrates, pigments, and antioxidants that can be converted into edible fruits, vegetables, etc. Use of this kind will make the "mush" into meals. Researchers have added milled mealworm to a shortbread cookie recipe. Future grocery stores are going to stock "food cartridges" thereby freeing up shelf space and making the unnecessary transport and stockkeeping needs. This research has been a center force of attraction in the case of meat substitutes. A few researchers have been trying algae as an alternative of animal protein, whereas certain members are researching to make animal meat using cow cells.

9.6.2 Internet of Things (IoT)

Digital technology has been transforming the agricultural industry, whereby making advances to tackle the present-day challenges. The Internet of Things and its technology based on correlation techniques used in machine learning thereby allows the diagnosis across three different kinds of data and thereby making future business insights for the production of food. IBM's Watson used an IoT platform by utilizing machine learning and artificial intelligence and thereby capturing sensor and drone data, making a transformation in the management of the farming industry.

9.6.3 Automation of Skills and Workforce

The United Nations has projected a two-thirds increase in the population of the world, which leads to the destruction of the rural world. In order to reduce the load of farmers these new technologies have enabled them to operate remotely by means of automation and thereby making life easier. The mix of modern biological farming practices in collaboration with the advanced technology makes reliable precision farming.

9.6.4 Data-Driven Farming

Utilization of sensor and transducer data viz., climate, types of seeds, quality of mud, disease prediction, old data, modern market trends, costs inclined can be used to perform a deep study, relating thereby making prior decisions by increasing their profit.

9.6.5 Chatbots

Artificial intelligence powered virtual assistants, chatbots can be used in retail, transport, media, and insurance industries. These technologies will influence farmers with high advantages over specified challenges.

9.6.6 Drone Technology

Drone-based technology has attracted more than $127 billion investment. Different ways with how this can be used are addressed as follows:

- **Soil and field analysis:** In order to facilitate this, 3D maps can be utilized for soil analysis, in planting seeds, and in analyzing and absorbing the data for irrigation, watering, and in finding the nitrogen levels needed for the growth of the plants.
- **Planting:** Several different start-up companies have enabled the drone planting equipment and making a decrease in the use of the cost of planting by 85%. These tools enable pods with seeds and nutrients into the mud, thereby providing the soil with all the nutrients needed for the growth of the plants.
- **Crop spraying:** Drones by imaging technique scan the crop field in real time and make the spraying of weeds and pesticides fivefold more than the traditional way.
- **Crop monitoring:** To monitor the crops by making dashboards, animations predict the crop development and production inefficiencies by making better management practices.
- **Irrigation:** To check whether the field is dry or not, these drones are more efficient than the human-based irrigation for farming.
- **Health assessment:** By scanning a crop using both visible and near-infrared light, a drone carries devices that can help track changes in plants and indicate their health and alert farmers to disease.

9.6.7 Unmanned Aerial Vehicles

Unmanned aerial vehicles (UAVs) coagulate autonomously and swarm across the field by means of collecting the data and thereby performing informed decisions. The most important is that the high-quality data number crunching software can make that high-tech dream a reality.

9.6.8 Blockchain and Securing the Agriculture Value Chain

Blockchain is a new distributed database ledger used for huge secure digital transactions and storage. This is the technology behind Bitcoin and other crypto-based transactions. Since it is a virtual currency, it can also be utilized for agriculture too. Blockchain can reduce inefficiencies and fraud and improve food safety, farmer pay, and transaction times. By improving traceability in supply chains, it can enable regulators to quickly identify the source of contaminated foods and determine the scope of affected products during contamination incidents. Additionally, the technology can reduce waste by detecting bottlenecks in the supply chain contributing to food spoilage. The transparency of blockchain can also help fight food fraud. As consumer demand for organic, GMO-, and antibiotic-free food soars, the news is rife with cases of fraudulent labeling. The smallest transactions, whether at the farm, warehouse, or factory can be monitored efficiently and communicated across the entire supply chain when paired with IoT technologies, such as sensors and RFID tags. Maersk, a shipping and logistics company, has intra-continent supply chains that involve dozens of personnel and hundreds of interactions. They estimate that blockchain could save those billions by improving efficiencies that reduce fraud and human error. The benefits of openness extend to all honest market participants. Blockchain technologies can prevent price extortion and delayed payments while simultaneously eliminating middlemen and lowering transaction fees, leading to fairer pricing and helping smallholder farmers capture a larger part of their crop value.

9.6.9 Nanotechnology and Precision Agriculture

Besides the green revolution in 2020, which was moved by the unprecedented use of pesticides and chemical fertilizers, thereby pertaining to compensate the loss of biodiversity with the resistive nature against pests and pathogens. Besides, nano-farming is going to revolutionize the farming industry to be more advanced than ever before. Nano-particles can be used for planting and advancing biosensors for precision agriculture. Nano-encapsulated conventional fertilizers, pesticides, and herbicides will release nutrients and agrochemicals in a slow and sustained manner, resulting in precise dosage to the plants. Among the benefits of nanotechnology precision farming are: 1. Roughly 60% of applied fertilizers are lost to the environment, causing pollution. 2. Nano-fertilizers help in the slow, sustained release of agrochemicals, resulting in precise dosages. 3. Greater plant protection and treatment of diseases. 4. Biosensors can detect pesticides in crops, leading to more-informed decisions.

9.6.10 Food Sharing and Crowd Farming

Economy sharing and crowd-based farming take place in reducing the waste of food. Several technologies have been made to deliver different goods and services. Most popular farming practice is the house farming or house sharing. Olio, founded by social entrepreneurs, has built an application that connects people and shops where the excess amount of food can be shared rather being wasted. Another social entrepreneurial project, Naranjas del Carmen, has developed the concept of crowd farming. Naranjas Del Carmen has created a system in which the person has ownership over the trees and land that the farmer cultivates. In this way, the fruit of those trees goes to their owners, creating a direct link between production and consumption and avoiding overproduction and waste along the value chain.

9.6.11 Internet of Agro-Drone Things (IoDT)

The do-it-yourself (DIY) approach is to build your own agricultural drone. Utilization of cloud computing and Internet of Things features are used for achieving this task. The quadcopter autonomously flies and captures images of the field, which are processed to find the condition of the crops. This image and corresponding result is stored on the cloud. The on-ground sensors and sensors present on the quadcopter help to monitor the temperature and humidity for better yields. By leveraging usage of the cloud, users can have ubiquitous access to remotely monitored farm data.

9.7 CONCLUSION

The agricultural drones play an important role in agriculture due to their remote access. It is easy to handle by any person. The data collected by these unmanned aerial vehicles improves the agricultural practices and regulates the processes. They support enabling precision agriculture that not only improves the quality and quantity of the product but also increases the revenues for farmers. This further has an impact over the country's revenue, which tends to increase thereby boosts the agricultural field. Thus, agricultural drones become part of agriculture throughout the world.

REFERENCES

[1] U. M. Rao Mogili and B B V L Deepak, "Review on Application of Drone Systems in Precision Agriculture", *Procedia Computer Science*, Vol. 133, pp. 502–509, 2018.

[2] Morey N. S., Mehere P. N., and Hedaoo K. "Agriculture Drone for Fertilizers and Pesticides Spraying", *International Journal for Engineering Applications and Technology*, Vol. 3, No. 5, pp. 1–7, 2017.

[3] Vanitha N., Vinodhini V., and Rekha S., "A Study on Agriculture UAV for Identifying the Plant Damage after Plantation", *International Journal of Engineering and Management Research (IJEMR)*, Vol. 6, No. 6, pp. 310–313, 2016.

[4] Primicerio J., Di Gennaro S. F., Fiorillo E., Genesio L., Lugato E., Matese A., and Vaccari F. P., "A Flexible Unmanned Aerial Vehicle for Precision Agriculture", *Precision Agriculture*, Vol. 13, No. 4, pp. 517–523, 2012.

[5] Reinecke M. and Prinsloo T., "The influence of drone monitoring on crop health and harvest size", 1st International Conference on Next Generation Computing Applications, pp. 5–10, 2017.

[6] Puri V., Nayyar A., and Linesh Raja, "Agriculture Drones: A Modern Breakthrough in Precision Agriculture," *Journal of Statistics and Management Systems*, Vol. 20, No. 4, pp. 507–518, 2017.
[7] Stehr N. J., "Drones: The Newest Technology for Precision Agriculture", *Natural Sciences Education*, Vol. 44, No. 1, pp. 89–91, 2015.
[8] Dutta P. K. and Mitra S., "Application of Agricultural Drones and IoT to Understand Food Supply Chain During Post COVID-19", *Agricultural Informatics: Automation Using the IoT and Machine Learning,* Wiley, pp. 67–87, 2021. 10.1002/9781119769231.ch4
[9] Reyes Peralta J. H., "Green Economy In Mexico Derived From The Use Of Agricultural Drones", Thesis, pp. 1–30, 2021.
[10] Tang Q., Zhang R. R., Chen L. P., Deng W., Xu M., Xu G., Li L. L., and Hewitt A., "Numerical Simulation of the Downwash Flow Field and Droplet Movement from an Unmanned Helicopter for Crop Spraying", *Computers and Electronics in Agriculture*, Vol. 174, pp. 1–14, 2020.
[11] Guo Q., Zhu Y., Tang Y., Hou C., He Y., Zhuang J., Zheng Y., and Lou S., "CFD Simulation and Experimental Verification of the Spatial and Temporal Distributions of the Downwash Airflow of a Quad-rotor Agricultural UAV in Hover", *Computers and Electronics in Agriculture*, Vol. 172, 2020.
[12] Zhang P., Guo Z., Ullah S., Melagraki G., Afantitis A., and Lynch I., "Nanotechnology and Artificial Intelligence to Enable Sustainable and Precision Agriculture", *Nature Plants*, Vol. 7, pp. 864–876, 2021.
[13] Yin H., Cao Y., Marelli B., Zeng X., Mason A. J., and Cao, C., "Soil Sensors and Plant Wearables for Smart and Precision Agriculture". *Advanced Materials*, Vol. 33, No. 20, pp. 1–24, 2021.
[14] Kashyap P. K., Kumar S., Jaiswal A., Prasad M., and Gandomi A. H., "Towards Precision Agriculture: IoT-Enabled Intelligent Irrigation Systems Using Deep Learning Neural Network", *IEEE Sensors Journal*, Vol. 21, No. 16, pp. 17479–17491, 2021.

10 IoD-Enabled Swarm of Drones for Air Space Control

J. Bruce Ralphin Rose, T. Arulmozhinathan, V.T. Gopinathan, and J.V. Bibal Benifa

CONTENTS

10.1 INTRODUCTION

The term *Internet of drones (IoD)* is a combined application of IoT and drones. Drones are technically known as UAVs (unmanned aerial vehicles) and a swarm or group of drones is called UAS (unmanned aerial dystem). Typically, the drones are unmanned and can be operated aerially as well as underwater, of course; both of them

DOI: 10.1201/9781003252085-10

have different purposes and different techniques involved in diverse operating conditions. In any case, since the drones are unmanned, they have to be operated remotely either manually or autonomously. Because of the rapid increase of drone applications in recent years, there is a steep increase in demand for drone pilots in the industry. Hence, the concept of Internet of Things (IoT) came into play to diminish the manpower requirements to the state-of-the-art autonomous flight modules with cloud computing and big data analytics. Therefore, the coordinated operation of drones and data access without latency have enabled numerous advantages in the advent of Internet of Drones (IoD). A schematic representation of the workings of UAV swarming with their data networks and cloud access is shown in Figure 10.1.

Generally, drones are classified based on their size, range, and the application equipment incorporated with it. According to the applications, the drones are equipped with cameras, warheads, sensors, global positioning system (GPS), medical aid kit, and many more sophisticated systems based on their specific requirement [1]. The size of the drones also varies as small, regular, and large based on their range and endurance along with the payload capacity. Moreover, the drones used for defense applications can be classified into two comprehensive types based on the geometry as follows: i). Fixed-wing UAV; ii). multirotor UAV.

The fixed-wing UAV is operated with a combined operation of manual gliding and automation that requires a separate launching mechanism. It is manufactured in a large size due to the requirement of its stability and less complexity during intense maneuvers. The multirotor UAV has a defined number of sets of rotor blades that are operated with tiny electric motors. The number of rotors in a drone may vary from three, four, six, or eight; thus naming them as tricopters, quadcopters, hexacopters, and octocopters, respectively. The quadcopter drone is often used for many domestic applications because of the simple construction and cost-effective nature. Alternatively, the multirotor drones exhibit less cruising speeds and they are operated for short-range applications as compared to the fixed-wing UAVs. With

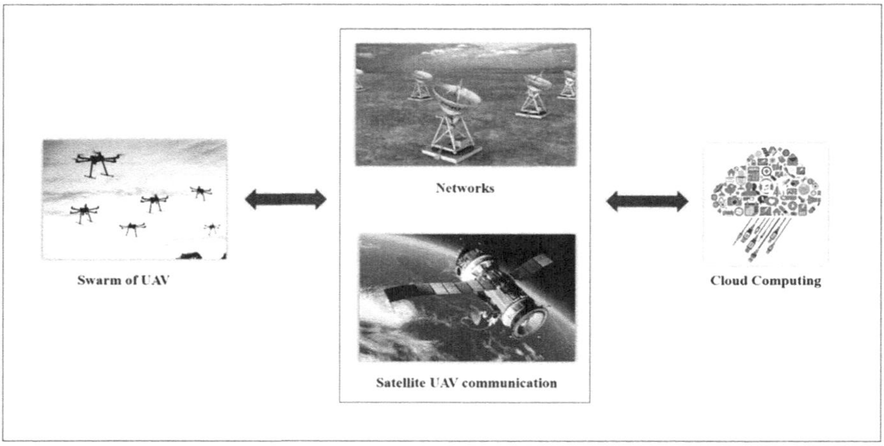

FIGURE 10.1 Representation of enhanced drone air space mobility using IoD (Courtesy: www.bing.com).

the advancements of aerospace engineering, the applications of drones have become immense and continue to increase across various disciplines. The concept of IoD has enabled drones towards defense and civilian applications with greater autonomy. A few selected defense applications of drones are summarized herein to emphasize the importance of IoD in swarming and better air space control.

10.1.1 Rescue

Rescue in defense is one of the most important missions that is recognized as an essential service. The rescue operations may involve a stranded target at an unknown location. A swarm of drones is the latest way to implement the rescue operations [1]. As an example, a life buoy being dropped by a rescue drone is shown in Figure 10.2(a) through the remotely given instructions.

10.1.2 Surveillance

Surveillance is a crucial activity and monitoring the borders of a nation is the hectic day-to-day work for the defense personnel. A typical border of a country may have mountains, rivers, deserts, and highly dense forests. Surveillance at these borders is a difficult task for the border security forces (BSF) to impede the trespassing activities in both ground and the air space. It is also essential to classify the object that trespasses the borders whether it is a foreign enemy or an animal that belongs to that civilian area. Hence, the small surveillance drones loaded with IoT-enabled cameras (Figure 10.2(b)) would assist the surveillance activity effectively in all kinds of terrains.

10.1.3 Detection of Bogeys

In recent times, along with advanced surveillance activities, the drones are also able to differentiate the bogey from a normal day environment in the airspace.

FIGURE 10.2 (a) Rescue drone, (b) survelience drone, (c) swarm of drones for military activity, (d) drone carrying a warhead, (e) firefighting drones, and (f) drones deleivering a medical aids.

Figure 10.2(c) shows a group of soldiers engaged in a swarm preparation of fixed-wing drones for the detection of bogeys. A bogey is an unidentified object that is in the airspace subjected to unwanted harmless activity or simply an attack from the enemy. The recent advancements in drone technology have enabled the usage of HD Gimbal and mapping cameras or thermal imaging systems that can identify or classify the bogey and destroy it by subsequent attacks.

10.1.4 Warfare

The main aspect of drone usage in the battlefield is to reduce human fatality. The modern-day drones are not only efficient in attacks but also capable of differentiating the target from other civilian locations through advanced artificial intelligence (AI) modules incorporated with them. A fixed-wing drone with a warhead to perform the suppression of enemy air defence (SEAD) role is shown in Figure 10.2(d). It can also be customized for the radar target classification through machine learning (ML) techniques to destroy the enemy radars.

10.1.5 Disaster Management

In the case of an emergency situation like a natural calamity or domestic violence, drones come in handy to estimate the severity of disaster, identify the disaster-intensive locations, and provide necessary supplies to the soldiers locked up in the specific regions. On the other hand, it helps to identify the anti-social elements or any other terror activities during the violence and the legal actions can be initiated according to the exact field data. Another important on-ground action is the supply of essential medicines and food during the disaster management activities. Figure 10.2(f) shows a medical drone delivering medical aid and these drones could reduce the disaster response time by 50% as compared to manual search and analysis.

10.1.6 Firefighting Drones

Firefighting is one of the daring tasks performed by fire service personnel during disaster management activities. A remote-controlled drone (tethered/untethered) or an automated drone using a thermal imaging process with IR sensors could identify the intensity of the flame and counter it with proper fire extinguishing techniques [2]. As the temperatures may reach over 600 K, human access is impossible to identify the trapped people inside a building. The modern variable area nozzles (VANs) proposed for various farming activities can be tailored to handle the fire extinguishing agents in the forthcoming days. Thus, the fire hotspots are identified precisely by drones and the throw of fire extinguisher balls/agents could reduce the intensity of flames promptly with a small blast.

A typical firefighting activity of drones is shown in Figure 10.2(e), and it can be used for both elevated locations as well as underground places where a firefighting mission exists [3]. For all these stated missions and applications, the tracking and mapping of fire patterns, identification of hotspots, containment, and damage assessment activities are involved. Hence, IoD plays a crucial role in terms of smart control system, flight performance, and flight safety and data security to fulfill the core objectives of the missions.

10.2 SWARM OF DRONES

A large collection of drones organized in a manner for a particular or multiple applications can be called a swarm of drones. Swarms are configured based on their field applications and availability of technology. In a swarm, each drone has to be controlled intrinsically as well as it will also be controlled or control other drones present in the swarm [4]. The quality of service improves with IoD/cloud-enabled swarm of drones because of the rapid data exchange and analytics about a specific operation.

A swarm in nature is an amazing feature where a large number of insects, for instance bees or groups of fishes, birds, sheep, ants and many other creatures in nature sense an obstacle in their path or a predator or a prey and move accordingly by avoiding them or approaching them. They use their bodily senses to sense the surroundings and act according to it. They do not have a particular leader or guidance for such cognitive instructions under various circumstances. For example, scientists have already revealed that ants use pheromones to communicate and the colony activity is accomplished through the smell of pheromones. In similar situations, the communication in the swarm of drones is done by a radio-controlled wireless mesh network, where the drones work in the frequency range of 2.4 GHz to 5.8 GHz. A bio-inspired swarm of drones can resemble a flock of birds, which do assemble themselves under various configurations during their flight. The most inspired one for the large aircrafts is the V formation of birds by which they reduce the aerodynamic drag in a large amount. The flock avoids an obstacle using some other formations like linear, complex., etc. [5].

For a fixed-wing drone, the stability is judged by its wings and control surfaces known as the ailerons, elevators, and rudder. Typically, the control surfaces are configured similar to the fixed-wing aircraft. The pitching moment is controlled by the elevators and the rolling and yawing moments are controlled by the ailerons and rudder, respectively. Whereas for the multi-rotor drone, the directional, lateral, and longitudinal stability are all carried out but varying the high-precision control of rotor (propeller blades) speeds accordingly. The rotors are also the only source of lift and thrust in the system. The swarm can be organized in multiple ways based on the volume of drones and their applications [6].

Let us consider a single-layered simple configuration of the drone system, as shown in Figure 10.3. The system contains four drones among which drone A is the leader drone and other drones, B, C, and D, are the follower drones. In the presented system, the drones are made to communicate to the leader alone or to communicate among all the drones, as displayed in Figure 10.3. In the multiple-layered swarm configuration, the drone is connected in different layers (i.e., multiple number of simple layer drones). The drones are grouped into several batches under their leaders and each leader from the batch communicates not only to the follower drones from its own batch but also to the leader drones of the other batches. In this

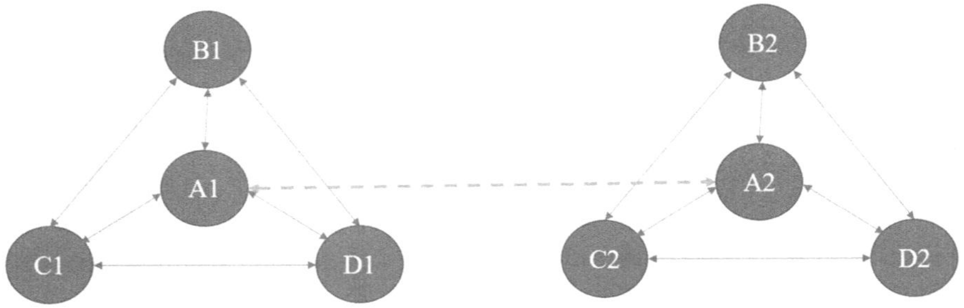

FIGURE 10.3 Two batches of drones in a single layer.

fashion, a large volume of drones can be implemented through a wireless mesh network for an application or multiple tasks [7].

A multi-layered batch of swarm is presented in Figure 10.4. It contains three batches of drones with A1, A2, and A3 as their leaders. The leaders communicate to their follower drones but also communicate to leaders of other batches. It can also be noted that batch 3 is at a different layer from the other two batches. Besides the given formation, any sizeable formation can be made possible. For instance, ring formation, linear formation, V formation, etc. [8].

Better swarming options under various missions at different weather conditions enhance the handling qualities and efficiency in the air space. Typically, drones are made to fly at a range of altitude from 30 m to 9,000 m depending on their design, application, and mission requirements. Hence, they undergo diverse weather conditions mainly depending on their flight altitude and geographic location. Unlike

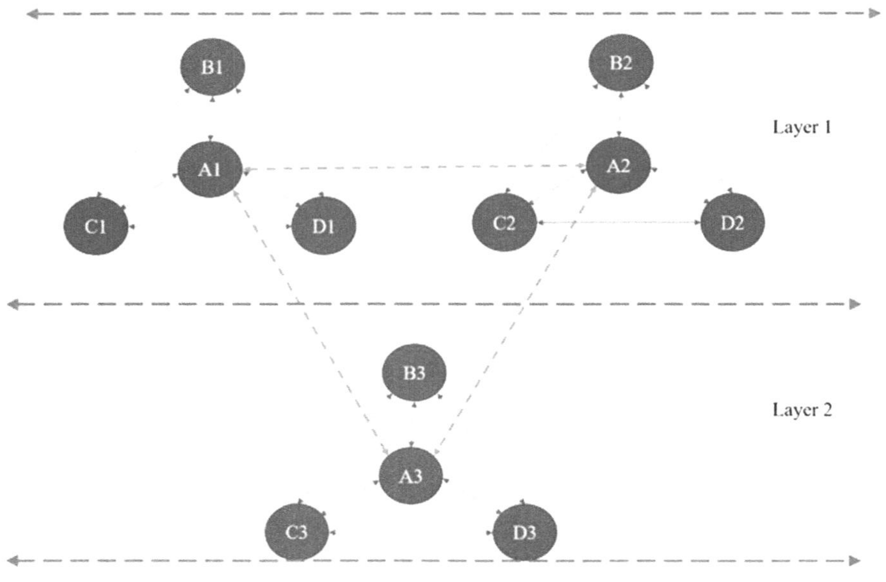

FIGURE 10.4 Multiple batches of drones at different layers.

large aircrafts, drones do not have the option of skipping the rain clouds by flying over them; thus, making it much more crucial for the drones to handle the situation. The possible factors that impact the swarm of drones in airspace are the high turbulent winds, gust encounters, torrential rains, and high temperatures.

For instance, a pass in the western Ghats, like Palghat pass or the Aralvaimozhi pass, present in the state of Tamilnadu, India, frequently have the high wind currents and isolated regions of Kerala and northeastern states in India have a very high rainfall almost anytime throughout the year. Also, if the temperature shoots up relatively more than 45 degrees, there could be uncertainties in the electronic data transfer that is expected in the state of Rajasthan and Odisha. Hence, a state-of-the-art IoD-based air space control is mandatory to enhance the functionality and handling qualities of swarms at specific tasks.

10.3 GENERAL SYSTEM ARCHITECTURE

The architecture for a swarm of drones requires three basic layers, as displayed in Figure 10.5. Firstly, within the swarm itself, the drones communicate among themselves and organize to process the tasks. They collect and communicate a large volume of data according to the computing capabilities available onboard. The second layer constitutes the network that transfers data from the drones to the cloud server/base station and vice versa [9]. The third layer consists of the cloud server that receives, computes, stores, and transmits the data to the drones as a two-way communication protocol [10].

10.3.1 Air Space Control

Identical to the normal day traffic for road vehicles, drones also undergo some unspecified air space constraints during their flight. Drones at their low-level altitude encounter many stationary and moving obstacles in their path such as terrains, towers, buildings, and probably another drone from the same network or a different one. In a swarm intelligence, two parameters play a vital role viz., collision

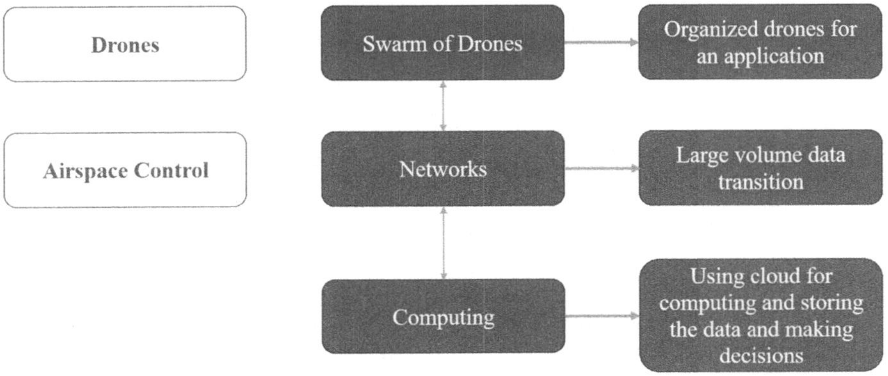

FIGURE 10.5 General system architecture.

avoidance and formation control. A few important examples of swarm intelligence are conceived from nature such as ants gathering food, prey birds escaping from their predator birds, and schooling of fishes. Specifically, the studies related to the V-formation of birds have revealed a significant amount of drag reduction, which is a nature-inspired phenomenon for better air space control [11].

10.3.1.1 Air Collision Avoidance

Collision – A sudden forceful direct contact of two bodies with an impact. Whenever the drone encounters an obstacle, it slows down to avoid collision using optimal sensors and takes a safe path. It is critical that braking in a vehicle causes a reduction in performance. Hence, with the advancements of today's technology, the equation of motion is synchronized with the forward and backward vision system (FBVS). The maneuvering characteristics of the swarm is customized through infrared, stereo vision, and ultrasonic sensors that work together with onboard GPS to set up enhanced performance at all operating conditions. It is essential to maintain the drone within the swarm during various maneuvers without colliding with its batch mates and also it should not go beyond the swarm boundary. Yasin et al. (2019) proposed an elaborated coordinate system that explains the actual position of the drone at any instant [12].

The coordinate system contains the following components:

a. Leader drone coordinates	– the position of the leader drone
b. Follower drone coordinates	– the position of the follower drones

Behavior of drones

a. Homing behavior	– to move towards a signal source
b. Cohesive behavior	– staying close to each other as a cluster.
c. Following behavior	– distinguish the leader drone from the batchmates and follow appropriately.
d. Dispersion	– to maintain minimum safe distance inside the batch
e. Alignment	– to move in a coordinated manner with respect to the attitude.
f. Conflict free behavior	– identify the location of batchmate drones and avoid conflicting their decision while avoiding an obstacle.
g. Boundary management	– with the aid of the homing signals, the follower drones locate the distance between them and their leader and maintain the required distance in the follow-up.

On a simple note, the swarm avoids a collision as a single swarm or splits itself into multiple batches with their respective leaders. Further, it splits the batches into individual drones based on the decision made to avoid the collision.

a. **Batch wise collision avoidance**
   The individual batches of drones remain in the same group for avoiding the obstacles.

b. **Individual collision avoidance**
   The individual drones disperse out from the batch and avoid the collision. They again rejoin the batch after dodging the obstacle.
   - In both cases, the stability of the drone is affected very much; thus, influencing the swarm formation of the drone system to perform the intended tasks.

c. **Narrow opening/path**
   Narrow openings are a special case that poses a serious challenge for the swarm concepts to identify the path size and align the drones in a sequence to use the available path. The critical decision to be made is that whether the drones are queued up one-by-one to pass through the narrow opening or by identifying the number of drones that can pass through the narrow opening without slowing down or queuing and simultaneously instructing the remaining drones to identify an alternate path to avoid the narrow opening while the initial set of drones travel through it.

Collision Avoidance Strategy

Collision avoidance is performed based on the three basic strategies listed as follows:

a. Optimization-based method is used to calculate the alternate path for the drones depending on the velocity, direction, and distance towards the obstacles by relying on the existing data of a static obstacle, their size, and dimensions.
b. Force-field-based method uses an attractive or repulsive electric field around each drone. When the drone senses another force field, it avoids it by moving away or stays within the boundary.
c. Sense-and-avoid-based method uses a sensor in multiple directions to identify the vicinity of the neighboring drone and maintains the distance to avoid collisions.

10.3.1.2 Formation Control

Formation control is generally achieved based on the three basic strategies:

a. Leader – Follower strategy

It is the basic mechanism adopted for the swarm of drones by following the leader and taking commands from it. Accordingly, it makes the drones align in the appropriate position using the coordinates referred by the leader drone. The individual leaders use a mapping technology or GPS to ensure their coordinates. The same shall be used for the follower drones or generate a sub-coordinate system with their leaders as their reference.

b. Behavior based strategy

One or multiple behaviors of the drones are combined together to maintain the formation between the drones. Commonly used behaviors are classified as cohesive, dispersion, and alignment behaviors. Selection of the specific combination of behaviors depends on the path or route of the swarm in which it is operated.

c. The virtual-structure-based strategy

The virtual-structure-based strategy would consider several drones as one unit of drone (i.e., a group of drones in the swarm as one drone). Thus, all the drones could reflect to a single command or decision simultaneously, irrespective of the number of layers.

10.3.1.3 Parameters Used for Formation Control and Collision Avoidance

A few vital parameters should be considered during the formation control and collision avoidance of the drones while implementing the stated strategies. It is required to maintain a safe distance between the drones and also it should be aware of the maximum clearance to move away from the neighborhood drones without breaking the formation. In order to avoid collision, the drones have to sense the distance between the obstacles and gather information about the size and shape of the obstacle. It is also important to identify the direction of the obstacle with respect to the direction of the swarm path. This input data is essential to determine an alternate path to avoid the obstacles without collision at any point of time.

A typical swarm of drones with an obstacle coordination system is presented in Figure 10.6. Here, the direction of the obstacle with reference to the direction of the motion is given as θ and ψ. The reference angles α, β, and γ are given to specify the direction of the obstacle as the swarm is in motion. The datum coordinate system (X, Y, Z) defines the distance between the swarm and the obstacle with respect to the reference frames [13]. A few parameters used for the collision avoidance studies are listed as follows:

D_{min}	– Distance to avoid collision
D_{max}	– Distance to avoid breaking up with the leader
D_{ob}	– Distance between the obstacle and drone
L_{ob}	– Size of the obstacle (based on its geometry)
θ, ψ	– Direction of the object
α, β, γ	– Angles made by the direction of the moving object in X, Y, Z axis, respectively

In the collision avoidance process, it is healthy to identify whether the obstacle is stationary or moving towards the swarm. Once when the size and attitude of the

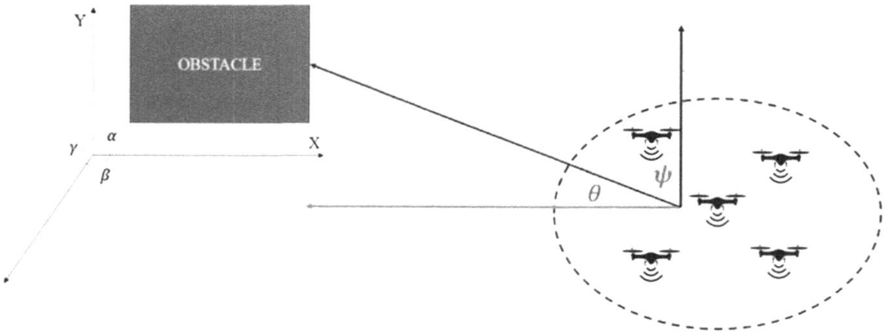

FIGURE 10.6 Swarm and obstacle coordination system.

obstacle is identified, the swarm takes the critical decision for avoiding the obstacle by various options, such as batchwise avoidance or individual avoidance or any other critical avoidance mechanism [14].

When a decision made by two different drones of the swarm is conflicting, by using the conflict-free behavior method the drones may take their own individual optimized path to avoid the collision with the obstacle. In such cases, if an individual drone happens to break away from the swarm, then it becomes a self-leader to progress towards the destination. Meanwhile, when it reaches the swarm boundary, it connects back to the swarm with the same leader or a different leader based on its configuration. It is also not necessary that a follower drone is dedicated only to a single leader. It can choose its leader based on the current situation of the mission and L-F formation control protocol involved [15]. The following algorithm is used for avoiding an obstacle that is stationary or moving towards the swarm.

i. Identify the obstacle: stationary or moving towards the swarm
ii. Choose alternate path based on the following data:
   a. Observe the dimensions and attitude of obstacles
   b. Single path or multiple path (forming sub-swarms to avoid the obstacles when multiple escape route is detected)
   c. Breaking away from leader and act as a self-leader to approach the destination
iii. Regroup as same swarm

In the outer environment, one more cases to be considered is the attacking attitude of the swarm. This shall be the sole purpose of a defense strategy during attacks. Similar to the collision avoidance system, the drones can arrange themselves to attack an unidentified bogey in the defense airspace.

In such cases, the initial step is to identify whether the bogey is moving towards or away from the swarm. If it is moving towards the swarm then the described algorithm can be used either to avoid or attack it. In order to attack a bogey, the path of the bogey has to be traced continuously and the chaotic behavior of the bogey should be adjudged (i.e., the next anticipated path of the bogey must be calculated and projected to the drones in the swarm) [14]. Based on that learning, the drones in the swarm would take multiple paths for encountering the bogey and trap it into a specific direction and launch the attack on the bogey [16].

The following is a simple step algorithm (steps (i) to viii) that can be used for encountering a bogey that is moving towards a swarm of drones.

i. Identify bogey: stationary or moving away
ii. Trace path of bogey (attitude of bogey)
iii. Adjudge the chaotic behavior of the bogey
iv. Split to multiple sub-swarm
v. Chase the bogey under the tail
vi. Trap the bogey from all possible directions
vii. Arrest the bogey motion to a specific target
viii. Attack and destroy bogey

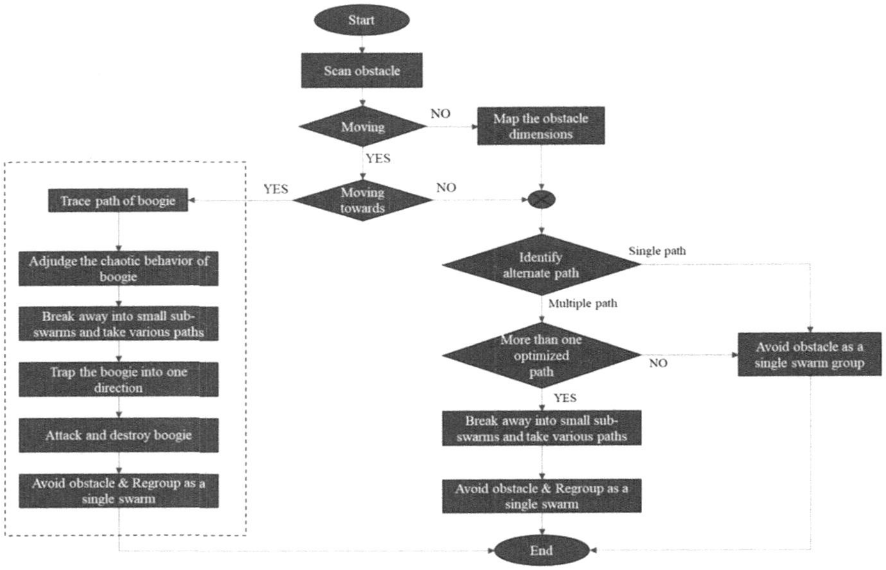

FIGURE 10.7 Flowchart for collision avoidance and bogey target system.

An illustrative flowchart for collision avoidance and bogey target system is presented in Figure 10.7. In a specific case, if the bogey is not to be attacked, then the swarm is enabled with the option to avoid the bogey instead of attacking it. The swarm of drones to avoid an obstacle can be tailored based on any bioinspired formation of birds, fish, ants, bees, and sheep that are available in nature. These abiotic engineering systems should be trained with appropriate machine learning (ML) techniques such as global-to-local safe autonomy synthesis (GLAS) to configure a greater number of formations that can be implemented on drones.

10.4 WIRELESS NETWORKS

In general, wireless mesh networks are established among drones for communication and the initial position of the leader drone can be located using GPS (global positioning system). The follower drones not only use the GPS but also use other sensors like infrared or ultrasonics to identify the position of other drones in the batch that allows the drone to avoid collisions and keep up the batch in the required position. Hence, the position of the leader drone is continuously monitored and in turn the leader drone monitors the follower drones through a multi-robot motion-planning algorithm.

In another method, a wireless communication is implemented in the form of a grid, where the coordinates of the air space is fed into the drones. The drone leaders identify their positions with respect to the allocated grid location. The method requires a mapping strategy to generate a grid that can be adapted by the drones. By this method, the position is fixed more accurately to the drones and collision avoidance is achieved almost every time. However, the pre-trained application cannot be changed as the mission of the swarm is very much pre-defined and it is not feasible to change during its operation. The only other way is to abort the drone's mission and this method is preferred for regular surveillance and fixed target missions. Alternatively, the GPS method described earlier can be used for any type of missions with fixed or changing targets requirements [11].

The drones move at a good speed in a given path that is frequently linear parabolic in nature with twisted curves. Hence, the communication system between the leader and follower drones uses 413 MHz of radio frequency, whereas the communication among the follower drones is established at 315 MHz frequency. Subsequently, the collected data can be transmitted and received from the cloud servers using the following networks.

10.4.1 Cellular Networks

The cellular network is one of the widespread networks in today's world. The same wireless network can be used to operate the drones in large volume. Similar to the air traffic control systems used for the common aircrafts, these cellular network towers can be used as air traffic towers. For instance, the first step is to segregate the network into multiple zones with a cluster of towers in a single zone. Each tower can be allotted as a separate sub-zone under its control. Once the zones are segregated, the drones are allotted for each zone Figure 10.8.

In the sub-zone Z1-T1, the drone is connected to tower T1. When a drone is traveling from one sub-zone to the other, say from Z1-T2 to Z1-T3, it is connected to both the towers in the overlapping area. After crossing the overlapping area, it will be totally disconnected from Z1-T2 and connected to Z1-T3. Thus, the same idea is implemented for drones traveling in between multiple zones, as illustrated in Figure 10.9. Since the drones travel into multiple zones and sub-zones, each and every drone in the swarm has to be assigned with a unique identification number (UIN) to avoid communication glitches. In this manner, a large number of drones can be connected to a swarm without any addition of battery-consuming electronic components [17].

10.4.1.1 Radio Frequency

Radio frequency has been the epicenter of many communication applications in the past century. With the influence of their former applications, UAVs also use radio

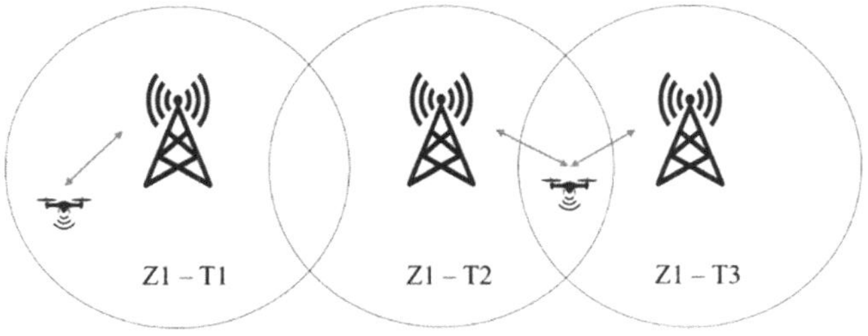

FIGURE 10.8 Single zone with multiple sub-zones.

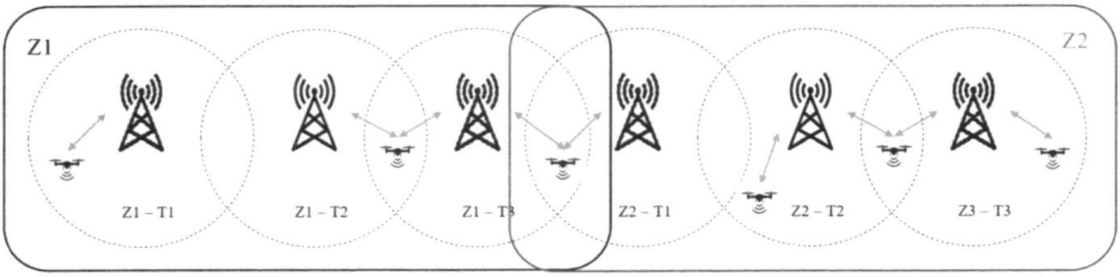

FIGURE 10.9 Two zones with multiple sub-zones.

frequency for their communication purposes. The radio frequency is used for UAV applications based on the consolidated information from the earlier research, experiments, and data that are available in the market [18]. Ground-to-air mode of transmission uses analog video signal with a reach up to a range of 5 km with a power of 6,000 mW. Analog transmitter signal uses a power of 2,000 mW to get a range of maximum 8 km, whereas digital video signal uses only a 1,500 mW. Long-range analog signals that have a frequency of 915 MHz use up to 2,000 mW to reach a maximum of 60 km. However, the EU standardization recognizes that for a maximum range of 60 km, a frequency band of 816 MHz uses a very little more than 1 mW. Air-to-air mode of transmission uses GPS and GLONASS with a frequency of 1,575.42 MHz and 1,602.45 MHz, respectively, with a range of 100 km.

The maximum range that can be reached by a particular frequency band is shown in Figure 10.10. Analog and digital video signal frequency are pretty low for video data but the transmitting signal frequencies are comparatively higher for GPS and GLONASS.

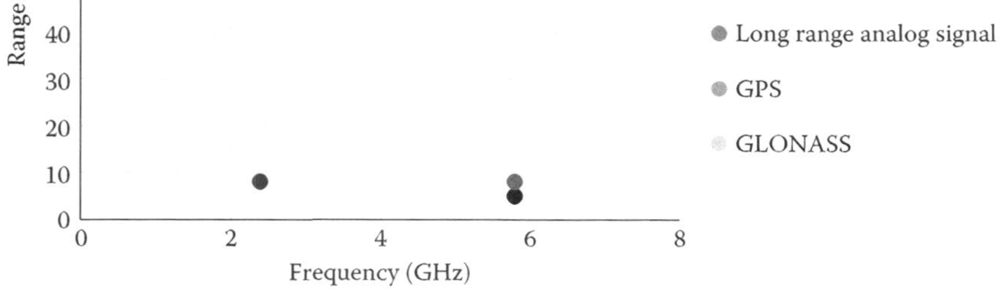

FIGURE 10.10 Comparison of rrequency and range for respective signals.

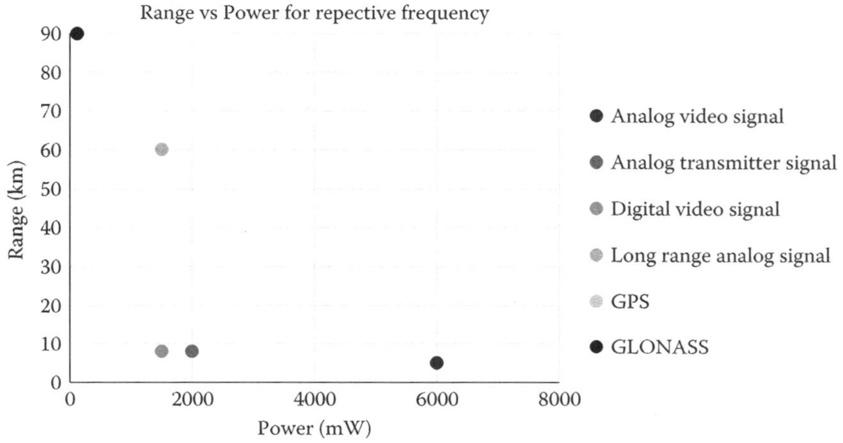

FIGURE 10.11 Comparison of range and power.

The power required to reach the maximum range for the particular frequency is shown as a scatter diagram in Figure 10.11. Short-range signals like analog video signals and transmitter signals use a low range of power, whereas digital video signals use relatively higher power. Until now, GPS and GLONASS have used a minimum power to achieve a long range [19]. However, the key insights presented in this overview of radio frequency are confined to a particular frequency that can be used for UAV applications. The comparison of different types of signals with their frequency are listed in Table 10.1 [20].

10.4.1.2 5G Networks

5G networks have an increased capacity, data transfer rate, decreased latency, and security to process the large volumes of data to execute a myriad of applications in IoT and cloud environments with big data analytics. 4G uses orthogonal multiple access that cannot be afforded by our applications since the time slots and bandwidth

TABLE 10.1
Comparison of the Type of Signal and its Frequency

S.No.	Type of Signal	Frequency	Bandwidth
1.	Map transmission signal	5.75 GHz–5.775 GHz	9 MHz
2.	Remote Signal	2.404 GHz–2.47 GHz	2 MHz
3.	WiFi Signal	5.15 GHz–5.85 GHz	20 MHz
4.	Leader-to-leader signal	413 MHz	–
5.	Leader-to-follower signal	315 MHz	–
6.	RADAR	3 GHz–30 GHz	–

allocated are insufficient. Hence, it is essential to note that today's technology with 5G networks uses higher frequency compared to 4G. As the number of IoT applications are rapidly increasing day by day, an affordable transmission of 1 Gbps data by a 4G network is a bottleneck and 4G is also more vulnerable to hackers and viruses. The data transmission of 4G networks is restricted up to 100 Mbps with the modern hardware and an upload speed of 50 Mbps only. However, 5G networks have the potential to transmit up to 1 Gbps data with 65,000 connections at a time. The key technology for 5G communication has created massive transformation in MIMO, millimeter wave technology, non-orthogonal multiple access transmission, device-to-device connection, and cognitive radio. The antenna in the UAVs could be integrated to the radio frequency for a massive multi-user MIMO [21]. By implementing the 5G, the overall cost of the UAV can be reduced along with its size and the heat produced by the electronic devices in the UAV also be reduced substantially [22].

10.4.1.3 Control Stations

Drones are used as network providing or enhancing devices in isolated places where the network is weak or overloaded with high traffic or even it can be deployed as a substitute for a base station. In such cases, drones act as a flying base station which will be mounted with a transmitter and receiver in such a way that numerous IoT-enabled devices are connected to it. These applications are known as drone-assisted communication, which will be a major support for defense technology and handling emergency situations like disaster management [21].

10.4.1.4 Radar Beams

Onboard ultralight radars are mounted on the drones for effective swarming. A MIMO transmit waveform is made by the drone signal to achieve a beam pattern using the drone-borne radar. This compact UAV-based radar system has been developed by the Lincoln laboratory with an active range of 15 km and it helps to classify the moving targets through raw data. The radar beam technology is used in places that are not GPS friendly. It also optimizes the battery consumption as compared to the individually operated radar-borne drone [23].

10.5 SATELLITE-UAV COMMUNICATION

UAV is primarily used for providing BVLOS (beyond visual line of sight) capabilities. Standard line-of-sight datalinks will be rendered unworkable at great distances due to the curvature of the earth, and drones may also fly out of range from the ground networks. SATCOM satellites are usually in geostationary orbit, where they appear at a constant position in the sky, or in low Earth orbit (LEO), where they appear to move. LEO satellites are at a lower altitude than geostationary satellites, which means that they will provide less latency, but geostationary satellites have the advantage of no tracking requirements. Satellite communications are split into different frequency bands. By increasing the order of frequency and decreasing order of wavelength, these are classified as L-band, S-band, C-band, X-band, Ku-band, and Ka-band. Higher-frequency bands usually provide greater bandwidth, but are also more susceptible to signal degradation because of signals being absorbed by atmospheric rain, snow or ice. The X-band is typically used by the military as a compromise between these two factors [24].

For a large swarm under a confidential mission, it can be connected to a purpose-oriented satellite with safe communication. It is the key component of communication in defense applications. The line-of-sight link is the most relied on for satellite-to-UAV channel. When using the Ka-band, it also suffers from rain attenuation. Since the location of a geosynchronous satellite is never changing relative to the Earth, the satellite-to-UAV communication prefers it. By considering the application and equipment used by the UAVs, the communication between the UAV and satellite can be done in different orbits during UAV navigation. The foundation for establishing a successful link for an UAV-to-satellite communication is the alignment of the spatial beam from the UAV to the target satellite. However, during continuous navigation, the UAV has to be continuously adjusted in the direction of the beam towards the target satellite to maintain the communication link [25].

10.6 CLOUD COMPUTING FOR DRONES

The major issue present in a swarm of drones is the transfer of big volumes of data with a limited energy supply. It should be noted that the drones are battery powered and the available energy is limited to perform all the computations onboard. The transfer of such high-volume data (i.e., the data received from camera, sensors for traffic, environment and specific application based sensors), could induce latency issues which is not acceptable for a swarm because the drones are operated on the basis of real-time decisions. Any latency in the data transfer could delay the crucial decision-making process of leader drones and it leads to collisions or other operational failures.

Ad-hoc cloud UAV swarms use the data between individual UAVs and form a virtual cloud and connect to an individual mobile cloudlet for their operation. The data latency in this method is low but the energy consumption is high even though the data resources are limited. Likewise, in a cloud-based UAV swarm model, the cloud robotics is used, thus outsourcing the onboard data computation to another data center. It causes high latency and high data consumption making it a difficult approach for real-time applications. Whereas in an UAV-enabled mobile edge computing model, the data latency is low and the data resource is high as compared to the cloud-based model. Therefore, to enhance the data capacity, a hybrid UAV-edge-cloud-based model can be used with AI systems to exploit valuable information. In this method, the drones are individually connected using a swarming method and the data is transferred to edge computing servers using a 4G/5G cellular network or mobile app network, as illustrated in Figure 10.12 [26]. Subsequently, the edge computing servers are connected to the cloud computing data centers [27].

A numerous amount of data has to be processed during the mission of a swarm, including the data for traveling and the data to complete the application-oriented mission. The mission planning component generates a deterministic plan by taking into account of the user input, available resources, and mission requirements. One such example of a mission is surveillance, where the time needed to cover the whole

FIGURE 10.12 Transmission of computed data between UAV swarms and cloud servers using various transmission devices.

area could be reduced depending on the desired image quality. It is done by choosing less overlap between neighboring pictures and using a higher flight altitude. Sensor data acquisition and analysis (SDAA) computes the overview images of high quality; it is important to choose the appropriate equipment. High-quality cameras are too heavy for small-scale drones. Lightweight cameras, on the other hand, are not as well developed and require essential setting parameters such as focus, exposure time, and white balance. Working with dozens of high-resolution images requires significant amounts of memory, computing power, and data rate. When mosaicking an overview image of large and structured areas taken from low altitude, it is important to minimize the stitching errors for every single image. State-of-the-art mosaicking tools fail in such cases, because the optimization goal is a visually appealing panorama from a single viewpoint. The multi-drone system has to deal with omni-present resource limitations on several dimensions.

The available onboard energy not only directly influences the total endurance but also it affects the payload and possible flight behavior and stability, especially in the windy conditions. Limited sensing, processing, and communication performance impede sophisticated onboard reasoning, such as performing real-time collision avoidance or online data analysis. Compensating a resource deficiency in one dimension often impairs another resource dimension. For example, flying at lower speed typically improves the image sensing but reduces the covered area. While the proposed centralized swarming approach allows for re-planning, a more adaptive coordination, where the drones decide their tasks on their own, would be beneficial, especially in case of dynamic environments. For instance, if the goal is beyond getting an overview image, e.g., tracking changes and dynamic events, the trajectories cannot be determined beforehand [28].

Recent studies have revealed significant information about the amount of data collected in proportional to the time consumed and speed of the drone in which it is traveling. Figure 10.13 offers one such comparison of amount of data collected with respect to time consumed at different drone velocities such as 5 m/s, 10 m/s, 20 m/s, and 30 m/s [29]. It is apparent that the time consumption increases for a fixed amount of data it the swarm of drones is subjected to a low-speed operation.

Each drone in the swarm collects individual observation and it is stored in the network attached storage (NAS) device in the cloud infrastructure. The NAS is equipped with a SPARK big data tool for managing the captured video. Each frame in the captured video is extracted and redundant frames are ignored. Then the

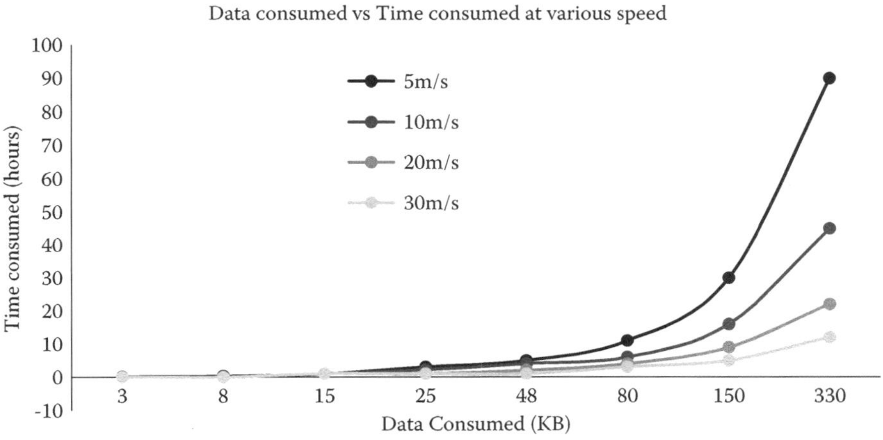

FIGURE 10.13 Comparison of amount of data collected, time consumed, and speed of drone.

extracted frames are passed to the computing server. Subsequently, the objects in each frame are identified by using you only look once (YOLO V3) object detection framework [30]. Then the cropped objects are subjected to the deep learning model GoogleNet that is trained over ImageNet dataset [31]. The GoogleNet model is deployed in the computing server and subsequent classification process is done. The SPARK tool supports the processing and management of big data collected by the swarm drones. YOLO is a real-time object detection algorithm that identifies specific objects in videos, live feeds, or images and YOLO was trained over COCO data set (Figure 10.14). YOLO pre-trained model is deployed in the computing server. YOLO uses features learned by a deep convolutional neural network to

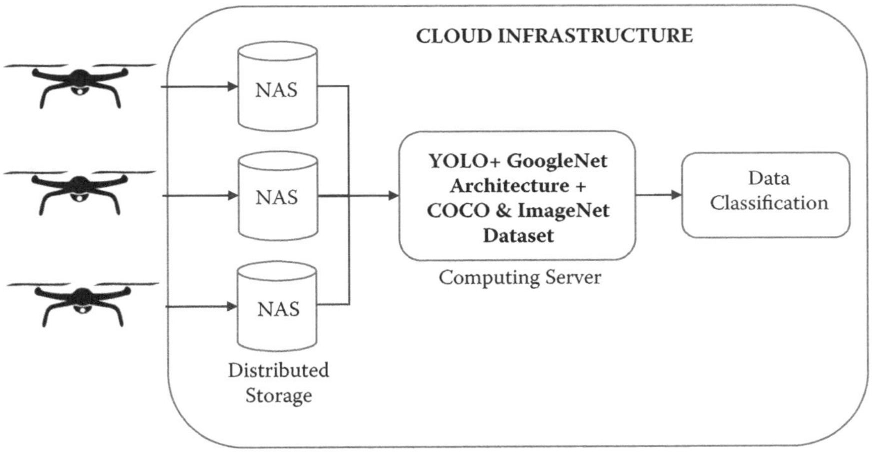

FIGURE 10.14 Drone data processing and classification in cloud infrastructure using deep learning model.

detect an object. GoogleNet is a 29-layer deep learning model that is trained over an ImageNet data set to classify thousands of real-world objects.

10.7 CONCLUSIONS

IoD devices are often connected with many sensors that generate a huge amount of data. As the 5G technology has higher bandwidth, such a high volume of data can be easily transmitted to the control devices or data centers wirelessly. IoT could also harness the advanced technologies used in 5G to work in an aligned and the most efficient manner. With advanced computing resources, IoT takes advantage towards swarm of drones and similar applications. However, in the future, a swarm network will be implemented for multiple applications at the same time, where each drone or each batch of drones will be following a different set of instructions to accomplish various missions under applications at the same time. Hence, a swarm would be incorporated with various designs of drones, different geometry, size, power capacity based on its application and mission. A quadcopter and fixed-wing drone shall be available together in the same swarm for different purposes at the same time.

In addition to the air space control and communication, some of the following complexities also should be considered while handling the high volume of drones, which certainly have an impact on the stealth nature in the defense:

- A considerable size of airspace is difficult for swarm camouflage.
- Noise suppression techniques.
- Larger network with high-speed data transmission and AI-enabled cloud support is needed.
- Human piloting should be nullified and more autonomous systems are needed.
- Large variety of autonomous algorithms are needed based on existing mission feedback.
- Security – strong lightweight cryptography is needed.

Since a large amount of data is generated by the IoD devices, the network that connects these drones must be stronger and fully encrypted. The power consumption monitoring also needs to be performed in real time as it becomes higher based on the application and mission of the swarm involved. Energy-efficient autonomous swarm systems with cloud-based congestion management could enable the collision avoidance with high rate of data transfers.

REFERENCES

[1] A. Tahir, J. Böling, M. H. Haghbayan, H. T. Toivonen, and J. Plosila, 'Swarms of unmanned aerial vehicles—A survey', *J. Ind. Inf. Integr.*, vol. 16, p. 100106, 2019, doi: 10.1016/j.jii.2019.100106.

[2] N. Jayapandian, 'Cloud Enabled Smart Firefighting Drone Using Internet of Things', *Proc. 2nd Int. Conf. Smart Syst. Inven. Technol. ICSSIT 2019*, pp. 1079–1083, 2019, doi: 10.1109/ICSSIT46314.2019.8987873.

[3] M. S. Innocente and P. Grasso, 'Swarms of autonomous drones self-organised to fight the spread of wildfires', *CEUR Workshop Proc.*, vol. 2146, pp. 30–39, 2018.
[4] M. Gharibi, R. Boutaba, and S. L. Waslander, 'Internet of Drones', *IEEE Access*, vol. 4, pp. 1148–1162, 2016, doi: 10.1109/ACCESS.2016.2537208.
[5] I. L. Bajec and F. H. Heppner, 'Organized flight in birds', *Anim. Behav.*, vol. 78, no. 4, pp. 777–789, 2009, doi: 10.1016/j.anbehav.2009.07.007.
[6] C. Kownacki and D. Oldziej, 'Fixed-wing UAVs flock control through cohesion and repulsion behaviours combined with a leadership', *Int. J. Adv. Robot. Syst.*, vol. 13, no. 1, pp. 1–10, 2016, doi: 10.5772/62249.
[7] G. Asaamoning, P. Mendes, D. Rosário, and E. Cerqueira, 'Drone swarms as networked control systems by integration of networking and computing', *Sensors*, vol. 21, no. 8, pp. 1–23, 2021, doi: 10.3390/s21082642.
[8] G. A. Dimock and M. S. Selig, 'Self-organization in bird flocks', *Sci. York*, vol. 18, no. January, pp. 1–9, 2003.
[9] A. Koubaa, B. Qureshi, M. F. Sriti, Y. Javed, and E. Tovar, 'A service-oriented Cloud-based management system for the Internet-of-Drones', *2017 IEEE Int. Conf. Auton. Robot Syst. Compet. ICARSC 2017*, no. Section 3, pp. 329–335, 2017, doi: 10.1109/ICARSC.2017.7964096.
[10] M. Campion, P. Ranganathan, and S. Faruque, 'A review and future directions of UAV swarm communication architectures', *IEEE Int. Conf. Electro Inf. Technol.*, vol. 2018-May, pp. 903–908, 2018, doi: 10.1109/EIT.2018.8500274.
[11] K. Loayza, P. Lucas, and E. Pelaez, 'A centralized control of movements using a collision avoidance algorithm for a swarm of autonomous agents', *2017 IEEE 2nd Ecuador Tech. Chapters Meet. ETCM 2017*, vol. 2017-Janua, pp. 1–6, 2018, doi: 10.1109/ETCM.2017.8247496.
[12] J. N. Yasin, M. H. Haghbayan, J. Heikkonen, H. Tenhunen, and J. Plosila, 'Formation Maintenance and Collision Avoidance in a Swarm of Drones', *ACM Int. Conf. Proceeding Ser.*, 2019, doi: 10.1145/3386164.3386176.
[13] X. He *et al.*, 'Multiobjective coordinated search algorithm for swarm of UAVs based on 3D-simplified virtual forced model', *Int. J. Syst. Sci.*, vol. 51, no. 14, pp. 2635–2652, 2020, doi: 10.1080/00207721.2020.1799110.
[14] J. N. Yasin *et al.*, 'Energy-efficient formation morphing for collision avoidance in a swarm of drones', *IEEE Access*, vol. 8, pp. 170681–170695, 2020, doi: 10.1109/ACCESS.2020.3024953.
[15] C. Zhuge, Y. Cai, and Z. Tang, 'A novel dynamic obstacle avoidance algorithm based on Collision time histogram', *Chinese J. Electron.*, vol. 26, no. 3, pp. 522–529, 2017, doi: 10.1049/cje.2017.01.008.
[16] J. Johnson, 'Artificial Intelligence, Drone Swarming and Escalation Risks in Future Warfare', *RUSI J.*, vol. 165, no. 2, pp. 26–36, 2020, doi: 10.1080/03071847.2020.1752026.
[17] T. Zeng, M. Mozaffari, O. Semiari, W. Saad, M. Bennis, and M. Debbah, 'Wireless communications and control for swarms of cellular-connected UAVs', *Conf. Rec. - Asilomar Conf. Signals, Syst. Comput.*, vol. 2018-Octob, pp. 719–723, 2019, doi: 10.1109/ACSSC.2018.8645472.
[18] https://www.scribd.com/document/475429742/Day3-1000-1100-Civil-UAV-monitoring-techniques-JiWeilin-pdf.
[19] J. Weilin, 'Civil UAV monitoring techniques', vol. 148, pp. 148–162, https://www.scribd.com/document/475429742/Day3-1000-1100-Civil-UAV-monitoring-techniques-JiWeilin-pdf.
[20] A. Hobbs, *Unmanned Aircraft Systems*. In Book: Human factors in aviation, pp. 505–531, Academic Press, 2010.

[21] B. Li, Z. Fei, and Y. Zhang, 'UAV communications for 5G and beyond: Recent advances and future trends', *IEEE Internet Things J.*, vol. 6, no. 2, pp. 2241–2263, 2019, doi: 10.1109/JIOT.2018.2887086.

[22] J. M. Khurpade, D. Rao, and P. D. Sanghavi, 'A Survey on IOT and 5G Network', *2018 Int. Conf. Smart City Emerg. Technol. ICSCET 2018*, pp. 1–3, 2018, doi: 10.1109/ICSCET.2018.8537340.

[23] M. Alaee-Kerahroodi, K. V. Mishra, and B. M. R. Shankar, 'Radar Beampattern Design for a Drone Swarm', *Conf. Rec. - Asilomar Conf. Signals, Syst. Comput.*, vol. 2019-Novem, no. Cd, pp. 1416–1421, 2019, doi: 10.1109/IEEECONF44664.2019.9048820.

[24] https://www.unmannedsystemstechnology.com/category/supplier-directory/data-communications/satellite-receivers-satcom/

[25] J. Liu, Y. Shi, Z. M. Fadlullah, and N. Kato, 'Space-air-ground integrated network: A survey', *IEEE Commun. Surv. Tutorials*, vol. 20, no. 4, pp. 2714–2741, 2018, doi: 10.1109/COMST.2018.2841996.

[26] W. Chen, B. Liu, H. Huang, S. Guo, and Z. Zheng, 'When UAV Swarm Meets Edge-Cloud Computing: The QoS Perspective', *IEEE Netw.*, vol. 33, no. 2, pp. 36–43, 2019, doi: 10.1109/MNET.2019.1800222.

[27] O. Bekkouche, T. Taleb, and M. Bagaa, 'UAVs Traffic Control Based on Multi-Access Edge Computing', *2018 IEEE Glob. Commun. Conf. GLOBECOM 2018 - Proc.*, pp. 1–6, 2018, doi: 10.1109/GLOCOM.2018.8647421.

[28] N. Athanasis, M. Themistocleous, K. Kalabokidis, and C. Chatzitheodorou, *Big data analysis in uav surveillance for wildfire prevention and management*, vol. 341. Springer International Publishing, 2019.

[29] C. Wang, F. Ma, J. Yan, D. De, and S. K. Das, 'Efficient Aerial Data Collection with UAV in Large-Scale Wireless Sensor Networks', *Int. J. Distrib. Sens. Networks*, vol. 2015, 2015, doi: 10.1155/2015/286080.

[30] J. Redmon and A. Farhadi, 'YOLOv3: An incremental improvement'. (n.d.). arXiv.org. https://arxiv.org/abs/1804.02767.

[31] C. Szegedy, W. Liu, Y. Jia, P. Sermanet, S. Reed, D. Anguelov, D. Erhan, V. Vanhoucke, and A. Rabinovich. Going deeper with convolutions. *2015 IEEE Conference on Computer Vision and Pattern Recognition (CVPR)*. 2015, 10.1109/cvpr.2015.7298594.

11 Drones for Disaster Response and Management

Jose Anand, C. Aasish, S. Syam Narayanan, and R. Asad Ahmed

CONTENTS

11.1 INTRODUCTION

The immediate search and rescue operations after a catastrophic event like earthquake, tsunami, and storm are exceedingly difficult. Natural and man-made disasters will require this search and rescue operations [1]. An explosion created by a terrorist is an event where immediate actions must be taken to save and rescue people. In such situations, the presence of toxic gases, smoke, fire, and wreckage of buildings obstruct the human force from doing their search and rescue operations. The teams need to find the survivors and the injured before they die [2]. The search and rescue operators focus on extracting the survivors first. The main problem here is the obstructions involved will delay the human inputs and it also involves certain risks. This problem of risking human efforts to do search and rescue operations is done using autonomous drones [3]. Motivation is the key that drives a person or a team to produce innovative ideas and products. For our team, the motivation came from the struggle that we had during the flood in Chennai. The flood affected the entire city and people struggled to make

DOI: 10.1201/9781003252085-11

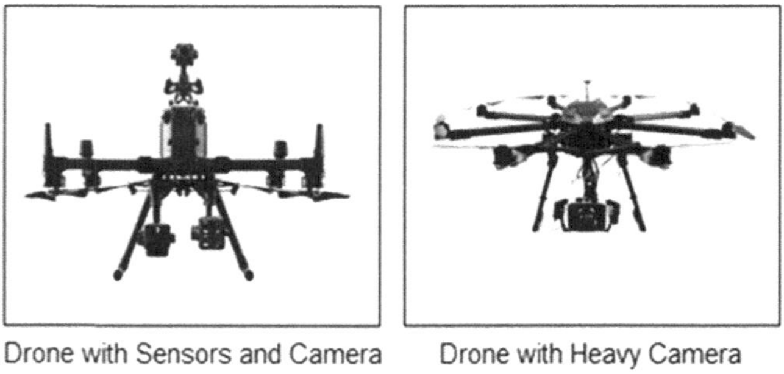

FIGURE 11.1 Drones with camera.

contact. Upon experiencing the struggle, the thought of the concept called "THE SAVIOUR," an aerial robot that can help people during calamities and can better be used in hostile situations by the soldiers. This technology can help people and assist the soldiers in planning to rescue hostages. Figure 11.1 shows the drones with sensors and lightweight cameras and normal cameras. The problem addressed in today's rescue operations is that a road network is used to access critical areas [4]. In certain cases, helicopters are used. Increasing congestion in the road network slows down the process. The search and rescue process is completely based on the human workforce. Critical situations in India highlight the slow response of emergency aid. Time is wasted by simply making arrangements. The problem of supplying medical aid is sometimes done by helicopters. This becomes expensive and it covers only limited areas.in other cases, only the human force is involved to do the required task, and personnel is made to take risks, do the search, and rescue operations. These existing methods have few drawbacks and limitations and involve time to operate.

GPS and to a lesser extent, simultaneous localization and mapping (SLAM) have become commonplace [5]. Some drones and unmanned aerial vehicles (UAVs) can hover using a navigation tool with a trained person in the absence of GPS or SLAM. Should UAV rely on navigation aids, particularly operating in a tight space? If done in a non-GPS environment with no clear physical indications, this chapter eliminates these off-the-shelf navigation methods. The optical flow approach is a navigation technique that could be applied. Honeybees [6] and other flying species have been shown in the lab to custom optical flow to determine velocity, height, and drift. There is no specific reason for not using any passive navigation options in UAV [7].

Advantages of Using Aerial System – The following are the advantages of using an aerial system:

i. Increased range
ii. Compact
iii. Fire and waterproof body
iv. Lightweight
v. Artificial intelligence
vi. Live telemetry
vii. Operation in sterile non GPS environment

viii. Two-mode navigation system
ix. Increased cruise and loiter capability
x. All-weather operations

11.1.1 Optical Flow

The deceptive visual gesture you perceive as you move over the globe is known as optical flow. Assume you're in a moving vehicle and watching out through the window. Plants, the ground, constructions, and other objects appear to move backward [8]. Optical flow is the term for this type of movement. This movement might also indicate how near on to the various substances. Clouds and mountains, for example, travel too lightly that seem to be stationary. Closer objects, like houses and plants, look to travel backward, with quicker substances moving quicker than farther substances. Substances that are remarkably close to you, such as grass or little road signs, move so quickly that they pass you by [9].

A clean mathematical relation among the value of the optic float and is related is observed. The velocity of the tour is doubled, the optic float you spot may also double. If an item is added two times as nearby, the optic float will once more duple ant its range relying upon the attitude among your route of the tour and the route looking at. If it is journeying ahead and the float is the quickest while the item is for the aspect through ninety degrees, or immediately above or under you. If the item is added in the direction of the ahead or backward route, the optic float might be much less [10]. An item immediately in the front of you'll don't have any optic float and seem to face still. However, due to the fact the rims of that ahead item aren't immediately beforehand of you, those boundaries are in motion and visualized as bigger. Figure 11.2 suggests what the optic float would appear to be from a plane flying above a rocky desert. Blue pointers display the optic float are visible through a digital digicam on the plane. Observing down, there may be a robust optic float sample because of the floor and rocks at the floor. The optic float is quickest immediately under the plane. It is especially rapid in which the big mainstay overhangs from the floor. A sensor at the plane that responds to optic float could be capable of seeing this optic float sample and apprehending the presence of the tall rock [11].

Looking ahead, there may be every other optic float sample because of the approaching rock and whatever else the plane is approaching. The blue circle immediately on the middle suggests the "Focus of Expansion" or FoE, which refers to the plane the unique route it's miles flying, if you're journeying in a direct line, the optic float is 0 within the immediately ahead route. Because of the big boulder on the left-hand side of this shot, the plane achieves a massive visual float to the proper FoE. Because of the ground, the plane also achieves slighter optic float styles in the downward-the-front path. It perceives no optic float in the top left of its field of vision because this area of the field of vision only has the sky. The plane will fly close to the enormous boulder, potentially dangerously close, according to the forward optic float sample. If the optic float at the plane's proper grows greater, the plane must interpret this as a signal to flee [12].

The advancement and application of UAVs in indoor environments have been increasingly popular in recent years. UAVs could be used to gather intelligence,

FIGURE 11.2 Optical flow observed from an aircraft.

conduct border patrols, and examine crowded settings to rescue people in the event of a disaster. Most UAVs are employed for reconnaissance. They need to be able to navigate to complete such assignments. In a crowded environment, GPS navigation is not possible [13]. As a result, an exceptional navigation approach is required. The navigation strategy observed in insects such as honey bees inspired me to choose optical float navigation as a GPS trading method. Navigation in sterile surroundings without outside navigation aids together with GPS or massive desk-bound factors of reference together with walls. The goal is to broaden a navigation gadget without the usage of outside navigational aids to rescue human existence at some stage in a disaster.

The goal is to develop a navigation system for UAVs that do not rely on external navigational assistance. The navigation system used is based on eyesight. The optical float is a sample of noticeable movement of objects, surfaces, and edges in a visual scene caused by the observer's movement relative to the scene. The optical float navigation approach is used for the mission, which is extremely useful in crowded or interior locations such as residences or city canyons. Utilization of hardware [14] and electricity intake within the optical float navigation may be very much less in comparison to different navigational aids. This navigation method may be very much like the navigation of honeybees and the alternative stated motives made it appropriate for me to select this method of navigation.

The chapter is prepared in any such manner that phase 2 highlights tremendous paintings within the regions of optical float navigation. Section 11.3, describes the methodology, running of sensors, cameras, and optical float. Section four describes the evaluation of the software program used within the gadget. Section 11.5 covers the modeling of SQUADRON 2 through the usage of layout software programs. Finally, phase 6 concluded with summary and destiny directions.

11.2 RELATED WORKS

Optical go with the drift has been extensively utilized by bugs and birds to help navigation functions. Such facts have attractive skills for the utility to small or micro-UAVs, especially for navigation and collision avoidance in the city or indoor regions. The reason for this bankruptcy is to offer a survey of present optical go with the drift strategies for UAV navigation packages. Detailed comparisons are made amongst exceptional optical go with the drift-aided navigation answers with an emphasis on the sensor hardware in addition to optical go with the drift movement fashions. The problem of navigation and management in GPS-deficient areas has been addressed by several writers [15]. Researchers have investigated scanning a range of sensors to map the surrounding environment and use this information for navigation.

In embodied categorization for imaginative and prescient-guided cellular robots, an opportunity method to imaginative and prescient-guided navigation eliminates imagination and prescient from the position of reconstruction absolutely [16]. This paper proposes a paradigm for a conceptual embodiment of creative and prescient-guided robots by laying forth a philosophical and psycho-physiological underpinning for embodied conception. The authors argued that categorization is essential in all degrees of robotic imagination and prescient. Further, classical laptop imaginative and prescient is mistaken for this categorization, however, thru conceptual embodiment, the energetic notion may be powerful. They gift a method for growing imaginative and prescient-directed robots that smears personification, obviously and indirectly, in categorizing visible facts to simplify green notions and achievement. The classical method has been the premise of tremendous improvement in laptop imagination and prescient is presented as a framework for breaking apart imaginative and prescient into a sequence of procedures that enabled segmental improvement with distinct regions of studies. The practicality of graded layers of dispensation has been confirmed through several imaginative and prescient structures, and researchers hold to make treasured contributions, especially regions. Modern-day laptop have imaginative and prescient algorithms recollect the perceiver to be an inactive unit and gives snapshots that needs to produce them with exceptional capacity [17]. Purposive imaginative and prescient, conscious imaginative and prescient, or energetic notion emphasizes that the connection between notion and the perceiver's physiology, in addition to the obligations finished, needs to be taken into consideration in constructing smart visible structures through authors. Rather, imaginative and prescient is used to achieve beneficial visible cues that could immediately offer a method of navigating within the surroundings, without using geometric fashions. In this manner, unique portions want now no longer be received; however, instead qualitative residences decided through the specific wishes of the project are extracted. Vision isn't usually considered in isolation; rather, it is embodied through a gadget. As a result, it is reason-driven and no longer necessary to resolve the broader rebuilding issue. In comparison to the traditional reconstructionist method, this provides clear computing advantages. The energetic idea is viewed as a searching, exploratory project in the paradigm of energetic notion. While the traditional notion has been built to serve the needs of navigation in the environment, the energetic notion reverses this dependency, causing actions to be inspired by a desire to understand. As a result, inventive and prescient work is dispersed and embedded in navigational obligations to shape observable behaviors, rather than being focused on maintaining a global version.

Removing the human from the loop in such situations could improve the efficiency of these activities [18,19]. Furthermore, completely self-sufficient robots that do not require any prior knowledge of their surroundings may be useful in a variety of fields, such as on dangerous manufacturing floors, in dangerous subterranean environments, or in transport yards where robots may also provide transportation services. There are distinct potential packages in which a person can be removed from dangerous situations and replaced by a robotic. To that goal, they developed a cellular robotic capable of exploring and mapping abandoned mines automatically. This robotic needs a durable electro-mechanical platform, a stable software program device, and a reliable technique of failure recovery to function without communications in difficult environments with little chance of rescue. The methods, algorithms, and evaluation equipment that enable self-sufficient mine exploration and mapping, as well as significant experimental impacts from 8 hit deployments, are presented. The self-sufficient cycle begins with a 3D experiment of the terrain ahead of the robotic, which is obtained by tilting the energetic laser scanner from 20 degrees above horizontal to 40 degrees below horizontal. This version is rendered as a 2 12 D terrain map, with the grey degree of each molecule within the map indicating its reversibility; lighter cells are easier to navigate than darker cells. They developed Groundhog, a cellular device for robot mapping of dry and partially submerged mines. Groundhog is a 700-kilogram custom-built platform with an inbuilt laptop, laser range finders, explosive fuel online sensors, and a low-intensity virtual visual device. It is controlled by a set of software program modules that handle everything from low-degree actuation to navigation and exploration. Groundhog can perform independently in subterranean voids, according to this article, which focuses on the device's concept, navigation, and exploration software program. Groundhog returns with information that can be used to create first-rate dimensional maps of its journey, as well as information that can be utilized to create a complete 3D of the mine. The results of an 8-hit Groundhog deployment into the derelict Mathies mine are presented in this paper. Groundhog has faced the hazards and demands of a desolate mine on eight trips and has completed seven of them without the need for bodily assistance. Groundhog has completed more than 10 hours of self-contained operation and has traveled more than a kilometer in a barren mine. Groundhog is a tough, trustworthy device that can gather information and create maps of dangerous places where no human should go.

The use of optical goes with the drift for imaginative and prescient-primarily based cellular robotic navigation, aiming to attain sturdy overall performance for navigational obligations [15]. There is convincing evidence of employing optical go with the drift in thought and navigation in animals in the biological imagination and vision. The visual movement has been identified as an important cue for navigational behaviors such as obstruction avoidance, grazing landings, focused UAV in corridors, and distance calculation in studies of imaginative and prescient flying bugs. Autonomous UAV is limited or cluttered environments together with homes or city canyons calls for excessive maneuverability, rapid mapping from sensors to actuators, and confined typical gadget weight. Although flying animals are nicely able to manage such situations, roboticists nevertheless have problems reproducing such skills. Authors took the notion from flying bugs to development towards the aim of growing small UAVs capable of dynamically flying in cluttered environments. This enterprise allowed us to illustrate a 10- gram micro flyer able for completely self-sufficient operation in an office-sized room the usage of fly-stimulated imaginative and prescient, inertial, and airspeed sensors. Regarding the sensor suite, they applied equal sensory modalities as in flies. Since omnidirectional imaginative and prescient isn't always viable on such light-weight UAVs, we opted for two extensive field of view (FoV), linear cameras. Only 3 segments of 20 pixels out of those cameras were decided on for optical go with the drift extraction in three directions: left, right, and down. Additionally, micro electromechanical systems (MEMS) gyros were installed to feel pitching and yawing rates. Motion detection and impediment avoidance have attracted excellent hobbies from laptop imaginative and prescient researchers because of their promising packages in regions, like robotics and site visitor monitoring. In a few situations, it's extra essential to grow the computation pace as opposed to optimizing code to grow the accuracy of the result. A magnificence of movement-making plans issues that have obtained extra interest is movement-making plans in dynamic environments with shifting barriers and shifting targets. It it proven that dynamic movement making plans for a factor within the aircraft, with bounded pace and arbitrary barriers, is intractable and NP-Hard. The optical go with the drift-primarily based totally approach of item detection is used to come across the item and those vectors are used to keep away from the impediment. This has stimulated new strategies for cellular robotic navigation the usage of optical to go with the drift. Regarding monocular imagination and prescient, a way exists that could offer facts the gap from a digital digicam constant on-board the robotic to an item, that's primarily based totally at the optical go with the drift vector subject. The optical go with the drift subject is received through processing or extra sequential perspectives of the equal scene. Such an approach has been explored in packages, like plant increase monitoring, video compression [20], automated automobile driving, and cellular robotic navigation.

The classical method to laptop imaginative and prescient is exemplified through Marr's concept at the position of imaginative and prescient. A string of procedures starts evolving with a fix of uncooked snapshots to provide an entire 3D version of the arena [21]. A strive is made to infer "an entire and correct illustration of the scene and residences" from the entered snapshots provided. The built version can then be utilized in next making plans and managing procedures to provide robotic motor responses. Next, psychophysical and computational investigations have questioned the significance of such unchanging 3D trends for reputation. In the last decade, computational imagination and prescience have been governed by a chance approach to reputation, which is centered on characterizing the possible photo appearances of an item rather than its invariant 3D structure. In a few strategies to laptop imaginative and prescient, the theory is of a trendy all-reason imaginative and prescient. He provides an issue from the philosophical literature that such a method is confined to what it can attain because of the wrong know-how of categorization. However, the philosophical idea of embodying categorization gives a manner ahead for laptop imaginative and prescient for robotic steering. Embodiment is the idea that humans have a physical foundation for their overall meaning, purpose, and creativity. This approach presupposes that a photo can be used to define the observed global. He used a paradigm to provide a strategy for developing a contemporary vision. This strategy excludes any incarnation or physiology for the gadget: it must accommodate all embodiments, purposes, and surroundings. As a result, either a single, uniquely accurate classification for all items in the globe is required, or creative or prescient is required to enumerate all descriptions and classes of seen items. This concept provides laptop imaginative and prescient a goal description of the arena that retailers may be visible inside a few organizations of the agent's studies community. While some aspects of reasoning can be considered independently of embodiment, he believes that when the agent is connected to a bodily world via laptop imaginative and prescient, embodied categorization must be the premise for a few portions of agent reasoning. Marr's classical theory is unsuitable for robotic steering because it requires that objects in the world be objectively subdivided into classes. As a result, a computer vision system must be capable of photographing a scene and generating a list of its contents without regard for the classification's purpose. This is no longer a problem of several algorithms, but of ways in which it is implemented. The traditional method has formed the foundation for significant advancements in computer imagination and vision. It provided a framework for breaking down imagination and prescient into a series of techniques that allowed for modular development with different areas of research. The value of hierarchical levels of processing has been demonstrated by various popular inventive and prophetic structures, and researchers, particularly in areas, continue to make valuable additions.

The layout and implementation of an actual-time extended Kalman filter (EKF), encompassing the layout of the unscented Kalman filter (UKF) [22]. The simulation effects and the experimental effects have been received through enforcing this set of rules in a visible-inertial sensor arrangement and the usage of a surprisingly green C++ implementation. The framework is for the estimation of the attitude, the pace, and the orthogonal distance to an aircraft. The clear-out is predicated on accelerometer and gyroscope facts for the propagation, and unmarried optical go with the drift capabilities for the dimension step. Simulations in Matlab affirm the derivation of the clear-out and supply an influence of its overall performance. With the addition of a powerful outlier rejection, the latter become included in the in-lab EKF framework multi-sensor fusion and become examined on actual measurements from the digital digicam and inertial measurement unit (IMU). In those experiments, it can be proven that the clear-out estimates converge efficaciously and in a surprisingly best way observe the full truth. They can show the overall performance in terms of robustness to outliers in the data, affiliation in the optical go with the drift tracker, and state initialization faults. It is proposed to use a UKF-based observer to determine both the total metric speed and distance to the aircraft, as well as its regular vector. The cited optical go with the drift-primarily based set of rules must be included in the flexible disbursed pose estimation framework. In this work, a modular EKF framework is offered, which makes use of IMU readings for its propagation steps. For the dimension steps, exceptional extra sensors may be used. Examples for such extra sensors are a Vicon gadget or a digital digicam, as within the case of optical go with the drift-primarily based totally on SLAM strategies. A digital digicam is used as a body-pace sensor within the aforementioned EKF framework. In this method, for the dimension step, the optical go with the drift among consecutive snapshots is extracted and derogated the usage of IMU readings. The scale is thought to be regular over a body and is represented through a metric scale aspect as a part of the kingdom vector. An optical go with the drift-primarily-based UKF for the kingdom estimation is offered. In comparison to, however, a tightly coupled method is chosen. This way, there isn't always the best one typical optical go with the drift fee among consecutive frames being fed into the clear out, however one optical go with the drift fee for each unmarried optical go with the drift characteristic itself. The estimator no longer relies on the aircraft's orientation, nor any distance or bearing sensors. Instead, the best optical go with the drift, and inertial sensors are utilized to enter the UKF as a dimension. Using this configuration, it is demonstrated in simulation that the clear-out device is now capable of estimating not only the aircraft's regular vector but also the entire attitude of the IMU-digital digicam gadget to the aircraft. They validate the used EKF, and the clear-out is reviewed on facts to estimate its overall performance. Facts are gathered by moving an ASL SLAM sensor with a synchronized digital digicam and IMU above an existing aircraft at a height of approximately 1.3 meters. After the recording, these facts should then be fed into the EKF. As floor truth, the output of a Vicon EKF becomes taken. This Vicon EKF makes use of IMU readings for 2 kHz for its prediction step and Vicon measurements at 200 Hz for its dimension steps.

Traditional optical go with the drift strategies with their extensions to sedation is conducted [23]. Optical go with the drift strategies the usage of gray-scale photo sequences is recognized to offer effective answers for movement estimation and form reconstruction packages. Advances in optical go with the drift computation from shading photo sequences are in comparison and delicate to offer opportunity optical go with the drift computation strategies for biomimetic imaginative and prescient and management structures. Thus, alternatively, there's a growing hobby amongst researchers in optical go with the drift techniques stimulated through the flying bugs as an answer for micro air vehicle (MAV) imaginative and prescient. Optical drift is the method through which flying insects with compound eyes, such as dragonflies, honeybees, or fruit flies, comprehend the environment. It allows organisms with little brains and low-decision eyes to gather enough information quickly to perform amazingly accurate navigation movements even in complex near-Earth surroundings. Without the use of GPS or radar for navigation, flying bugs must perform tasks such as collision avoidance, altitude management, and even take-off and landing. Such insect-stimulated navigational techniques can consequently function as a version for MAV flight styles in near-Earth environments. They looked at how the opposite researchers depicted how a MAV could experience optical drift while in flight in near-Earth environments, as well as how the optical drift would be utilized for small-scale navigation and collision avoidance. Vision could probably be the richest human sensing subsystem for the extensive gamma of facts it can offer. Then, to make the most of the imaginative and prescient gadgets to be had on-board, the cellular robotic is reasonable. However, for the sake of price reduction, the imaginative and prescient gadget often to be had on-board the robotic is a monocular one, and for that reason enforcing intense obstacles on a way to use imaginative and prescient to have beneficial facts encompassing the surroundings. Information approximates the range of items within the scene, for example, and will be generated through genuinely segmenting and obtaining the photo; however, the relative intensity of such items could now no longer be decided, for that reason making it not possible for the robotic to enforce the important maneuvers to keep away from them. Gray-scale optical go with the drift techniques has obtained an interest for getting better optical go with the drift in several packages. Color optical go with the drift, on the opposite hand, has now no longer been investigated as drastically and become no longer taken into consideration for UAV imaginative and prescient, no matter the supplementary facts this is to be had from the three channels of shading facts. Finally, they confirmed that using those shading optical go with the drift strategies should drastically beautify the overall performance of UAVs for the crowning glory of self-sufficient obligations and maneuvers like crash escaping, height supervision, take-off, and a touchdown in near-earth surroundings.

A method for self-sufficient robotic navigation is primarily based totally on the log-polar rework of snapshots and optical go with the drift [24]. The detection of impediments inside the traversable direction is part of the robot navigation project. This is regarded as a primary functionality for mobility, consisting of the dimension of the peak of items to categorize them as to be avoided or ignored. The FoE considers that the cellular robot moves with a translational speed parallel to the floor airplane, so the vanishing factor within the snapshot matches them. Navigation is primarily based on imaginative and prescient wishes to section the traversable direction and distinguish it from items that want to be prevented. The approach we endorse will clear up the trouble of impediment detection. The total performance of the picture assessment algorithms used is highly linked to the implementation of visible behaviors in synthetic structures. To meet the demands of the imaginative and prescient gadget activity, it's essential to broaden quick algorithms for photo processing and management, while displaying balance and robustness. The cortical mapping in the human visual system is accomplished using a space-variation sampling method, with the sample duration increasing linearly with the distance from the fovea. The sampling time will become virtually regular within the fovea. A metamorphosis from the retinal aircraft to the cortical aircraft (log (),) can be used to define this cortical mapping. This alteration is applied to the non-foveal portion of a retinal photograph. Log-polar mapping can be accomplished using ordinary photo sensors and remarkable space-variation sampling structures. The log-polar mapping has several important properties that make it a good sampling structure. The mapping of normal styles has an impact on other normal styles within the various areas. These mapping characteristics, along with rotation and scale invariance, are required for a few dwellings. Shifts along the c and c axes are caused by rotation and scaling, respectively. In the case of rotation, invariance recognizes that each of a factor's valid angular orientations at a given radius maps to the same vertical line. The projects for unstructured or outside environments and indoor environments of creative and prescient-based impediment detection can be divided into two categories. The former was solved using a sedation-comparison approach and 3D data was created using three independent creative and prescient modules based on brightness gradients, RGB (red, green, blue) sedation, and HSV (hue, saturation, value) sedation, respectively. They confirmed in this newsletter that it is feasible to widen a set of rules for obstruction avoidance based entirely on projective geometry throughout the test. The equations for the homography H were utilized by the authors to distinguish between the floor and non-floor aircraft capabilities. And they offered a method without fixing the equation explicitly however emphasizing that the capabilities flow alongside the epipolar traces defined. They used optical go with the drift to get the FoE and assigned the previous because of the significant factor of a log-polar rework. And additionally, they confirmed that the capabilities observe parallel traces within the log-polar area that may facilitate characteristic tracking.

This survey is bringing insightful facts on using optical go with the drift navigation strategies. As far as the facts affiliation trouble is concerned, it's far widely recognized that it absolutely depends on the capacity to locate the right capabilities and keypoint matching that's taken into consideration for a crucial project. In addition, using imaginative and prescient to outline landmarks appears to open an extensive variety of countless opportunities and the quandary that arises from the usage of cameras comes from the dynamism of the robotic surroundings. And this bankruptcy gives a complete survey of imaginative and prescient sensor hardware and reference movement fashions for the usage of optical flows to help UAV navigation. Several representing examples for optical-go with the drift-primarily based navigation of each constant-wing and rotary-wing UAVs have seemed inclusive of impediment avoidance, terrain following, vertical touchdown, pace estimation, and visible odometry. Although researchers have performed one or many of the above navigation functions, the maximum of the effects is nevertheless both confined to indoor environments or confined through computation powers. In addition, quantitative assessment and systematic calibration of present optical-go with the drift-primarily based navigation structures are wished earlier than actual-global packages.

11.3 METHODOLOGY

11.3.1 Aircraft Navigation and Path Planning

With the two-mode navigation system, all can navigate in the presence as well in the absence of GPS. It can map the area on its own and gives precise data about the survivors. All the 3D maps and data about the area, survivors, and other details are streamed live to the ground station. This navigation system installed makes it more robust. The frame is fabricated with fire and waterproof lightweight materials making the robot durable to operate in tough conditions.

As with any system that pushes forward and breaks records, there are obstructions in its path. Funding shortfalls function as a major barrier for this project. Apart from this, the import of certain components from foreign countries is often held by customs. The product's shipment delays the overall process of the project. Certification and legislation are other big barriers to this project. Due to frequent changes in the country's law, we are not able to do the testing of our prototypes. Time, legislation, and financial constraints are three major barriers to this project. To avoid customs, the procurement of components from local suppliers is sometimes too expensive or the materials will be below standard. A quality issue arises with the purchase of local components. This issue again acts as a major obstacle for this system.

11.3.2 Technology Readiness Level

The proof-of-concept model of the actual model is being made and tested by our team. All the basic functions of the robot are tested. Basic UAV tests were conducted separately for every sensor. The stability and control of the system were tested both autonomously and manually. We now wish to develop the full system integration on the actual model that can be used as a technical demonstrator. Further, this can be used for static display and live demonstrations. The system entails creating a navigation system for UAVs such as quadrotors that do not rely on external help. Quadrotor helicopters have four motors and are multi-rotor helicopters. The payload is mostly determined by the aircraft's total weight; hence, the overall system weight must be kept to a minimum. When associated with previous navigation schemes, this OFN requires fewer electronics and has a lighter overall system weight.

11.3.3 Process of OFN

OFN will be the best solution in these situations. The pattern of apparent motion of objects, surfaces, and edges caused by the relative motion of the observer and the scene is known as optical flow in a visual scene. A downward-facing camera is the main component of the OFN. The multi-rotor device that will be utilized to demonstrate OFN is a UAV. OFN will be the best solution in these situations. The pattern of apparent motion of objects, surfaces, and edges caused by the relative motion of the observer and the scene is known as optical flow in a visual scene. The process flow of OFN is represented in Figure 11.3. An automatic flight control system (AFCS) is included in the UAV, as well as a significant processing unit. A reference image is fed into the most important processor as an input. The photo is taken with the onboard digital digicam that shoots downward. After that, the acquired image is delivered to the CPU. After that, the processor compares the acquired snapshot to the reference photo. The working of the optical flow sensor is shown in Figure 11.4.

To perform the collection of operations, it is developed in the Python programming language. Python is a good programming language to use. It has environmentally friendly high-stage information systems as well as a simple yet effective object-oriented programming style. Python's stylish format and active keying, collectively with its inferred environment, make it a perfect linguist for

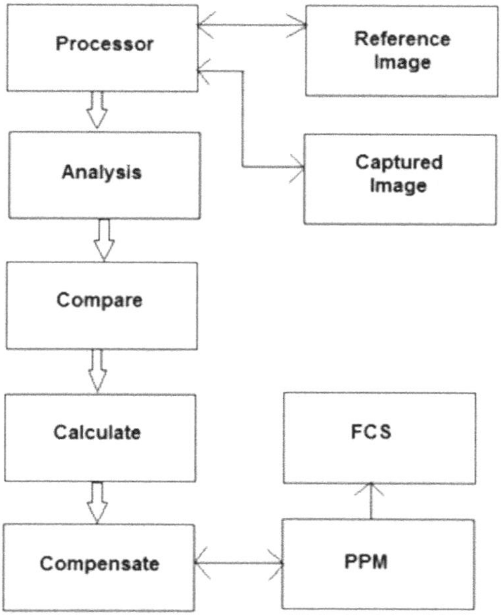

FIGURE 11.3 Process of OFN.

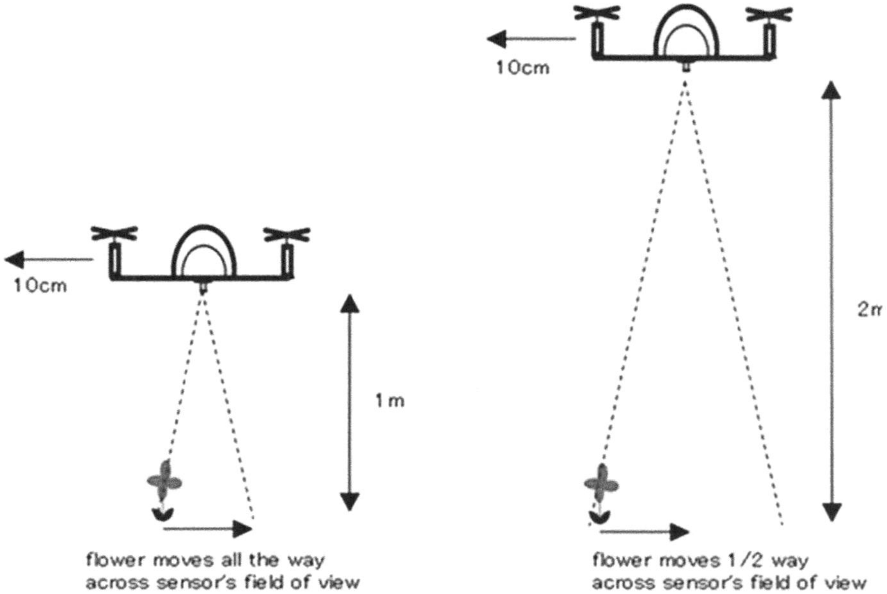

FIGURE 11.4 Working of optical flow sensor.

writing and fast software improvement in regions on maximum structures. The Python translator and the tremendous popular library are free to be had in supply or binary shape for all principal structures from the Python Internet page. The Python translator is without problems prolonged with novel capabilities and information sorts conducted in C or C++. The SURF set of rules for image comparison. The x and y coordinates are intended, and additionally, a blunder estimate is done. The remunerated standards are then transformed to pulse modulation alerts and nourished to the UAV to manipulate the device. The UAV manipulates the device and includes gyros and accelerometers and affords steadiness for the aircraft.

11.3.4 Optical Flow Field

Matlab code on phase-based technique is used in the optical flow field for publicly accessible. The process of computing 2D module speed was completed, and the findings revealed that stage outlines are additional resistant in the face of smooth shading and illumination fluctuations, as well as more stable in the face of modest deviations from picture translations. When flying in a congested area, adjust the UAV velocity to a lower level. For this, a simple approach can be devised. Reduce till the optic flow is high or the "pleasant" level or the optic flow around surroundings is increasing rapidly.

11.3.5 Collision Avoidance

Obstacle collisions are a major obstacle in navigating through congested surroundings. To navigate through congested areas, the UAVs will need an obstacle avoidance system. To avoid obstructions, a vision-based technique can be used. The task of vision-based obstacle identification can be divided into two categories: outdoor (unstructured) and inside (structured) situations. Visual flow navigation usage maintains the flight moving in crowded situations, but an obstacle avoidance system will make it more efficient. Furthermore, this navigation method consumes extraordinarily little energy and requires truly little equipment. As a result, using this navigation reduces the overall weight of the system, allowing it to be employed in ultra-light aircraft. The onboard computer with programs written using Python programing language processes the images to conduct the OFN. The sonar-based collision avoidance system integrated with the UAV controller enables safe operation and successfully conducts the mission.

11.3.6 Modeling Using SQUADRON 2

3D modeling is the procedure of creating a mathematical depiction of the 3D exterior of a substance using specialist software in 3D computer graphics. A 3D model is the product's name. It can be rendered as a two-dimensional image using a technique known as 3D rendering. CATIA was used to create the structural design for Squadron 2. With the ICEM surfacing technologies, this program explains shape planning, fashioning, sprouting process, and picturing to generate, edit, and test compound novel forms. CATIA can help you with all phases of product

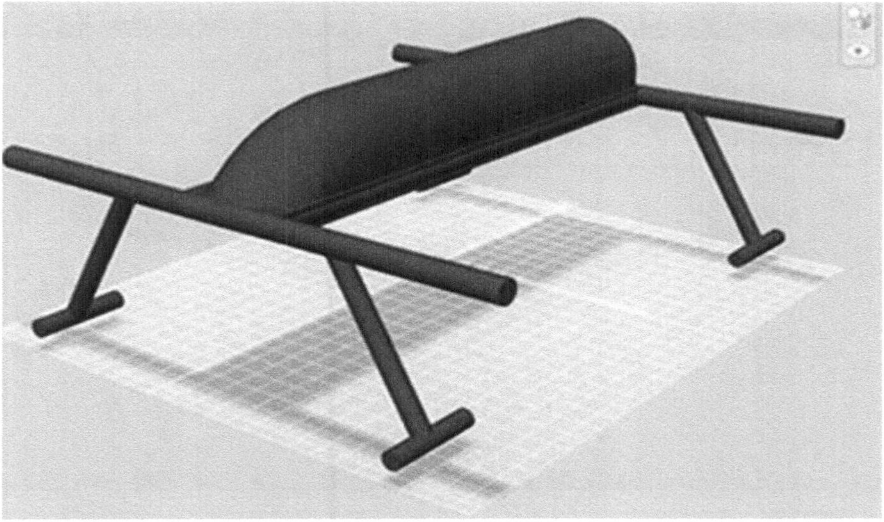

FIGURE 11.5 Overall structural design of SQUADRON 2.

development, whether you're starting from basic or using 2D sketches. For reverse engineering and surface reuse, CATIA can read and write STEP files.

The UAV platform chosen is a stable multi-rotor platform. This is a quadrotor platform named SQUADRON 2 that uses two pairs of rotors spinning clockwise and anticlockwise, respectively. The multi-rotor platform is chosen mainly because of its ability to hover. Figure 11.5 shows the 3D drawing of the multi-rotor platform. The structural diagram is drawn using 3D software. The 3D drawing helped us to place the payload on the structure in a convenient manner. The sequential arrangement of the payload was made only after this 3D drawing. The simulated version helped us to analyze the design before fabrication. In 3D graphics, 3D replicas are commonly employed. The extensive usage of 3D visuals on PC precedes their adoption. Before computers could create 3D objects in real time, PC playoffs use pre-defined 3D images as sprites.

3D models are being employed in a varied range of industries. They are used by the engineering community to create new technologies, cars, and structures, among other things. 3D models served as the foundation for real devices created with 3D printers and CNC equipment. The SQUADRON 2 frame is shaped like a letter I. This is different from how X and + type setups are normally designed. This design does away with the practice of stacking components one on top of the other, which is common in other layouts. A major benefit of this design is the ease of access to components and the payload system. Carbon fiber and 3D-printed elements are used to construct the SQUADRON 2. The complete system is designed to be lightweight while sturdy and stiff. CNC is used to cut the drawn designs. A propeller guard is meant to reduce the direct effect of the propellers on obstructions. The total system damage is also reduced by this protection. As a result, the quadrotor platform is built to be extremely stable.

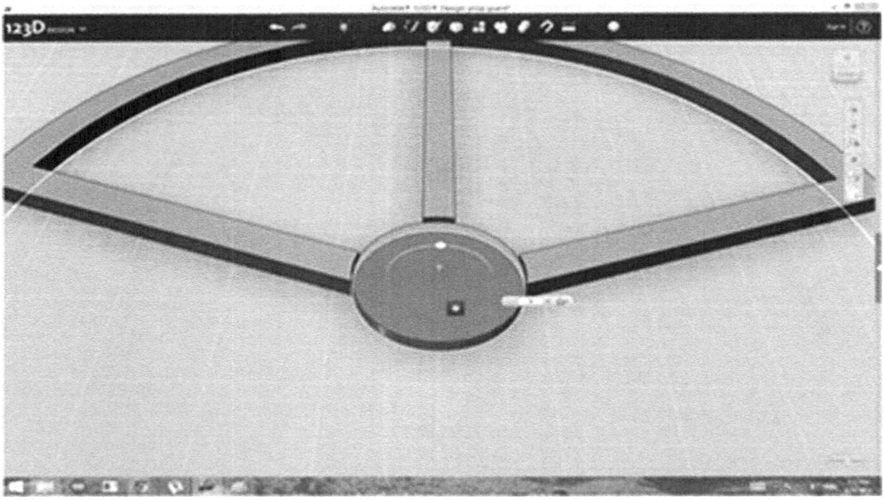

FIGURE 11.6 Design of propeller shield.

The propeller shield is designed using 3D software and is fabricated using a desktop 3D printer. The propeller shield is made of PLA plastic. We used the fused deposition method of printing for fabricating the propeller shield. The propeller shield protects the propeller of the UAV from getting damaged and thus making the overall design a robust design. Figure 11.6 shows only the 3D propeller designs that are then fed to the 3D printer for fabrication. Considered here are multiple designs for the propeller shield and the result in Figure 11.7.

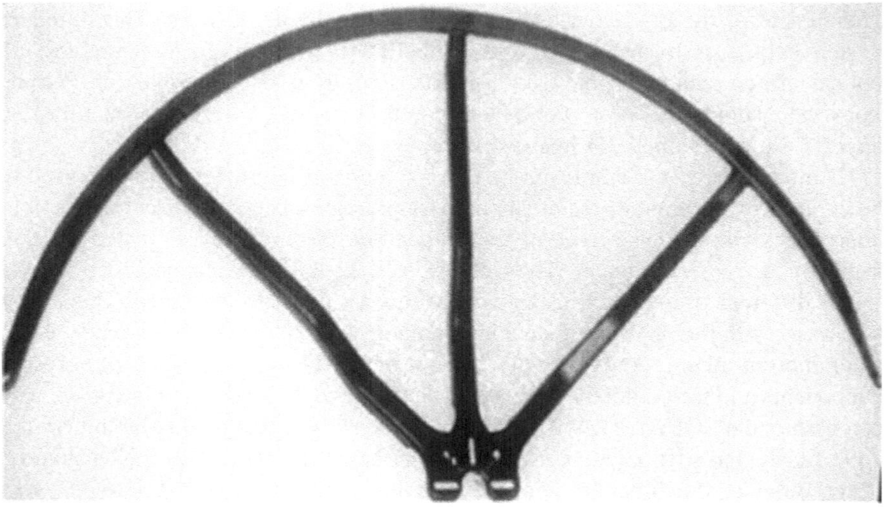

FIGURE 11.7 Fabricated propeller shield.

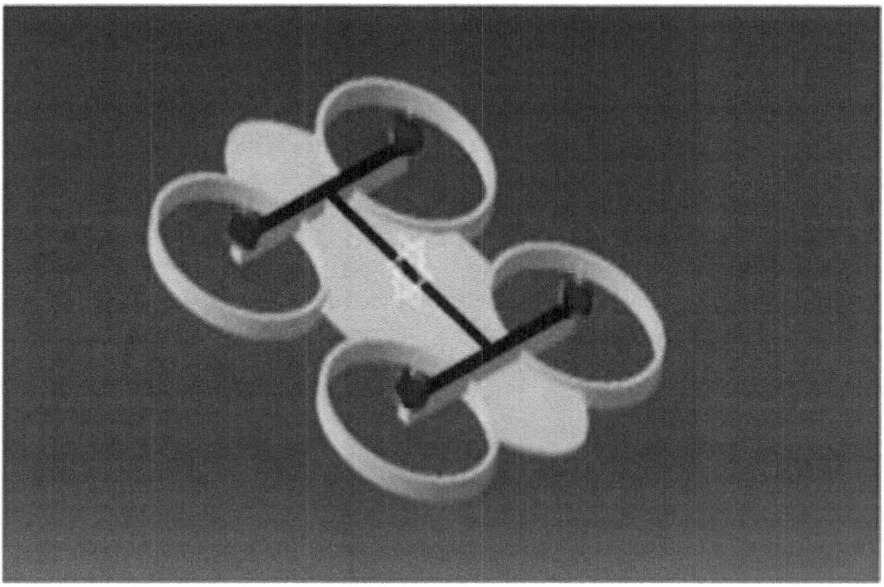

FIGURE 11.8 SQUADRON 2 design with propeller shield.

The propeller shield is designed in such a way that, there is much clearance between the tip of the propeller and the propeller shield. The design is made with a focus in mind that the thrust of the motor is not affected and is shown in Figure 11.8.

3D drafts through mixtures to the description of machine-driven assembly, the 3D modeling software facilitated the development of 3D parts. The software includes superior mechanical surfacing technologies. It includes tools for completing product definition, such as functional tolerances and kinematics with a wide range of tooling design presentations, including both generic tooling and mold and die design. This latest technology of rapid prototyping is a great boon. This technology made our work easier and helped us to do and evaluate frequent design changes. with this machine, we were able to fabricate the parts overnight and do changes immediately. The ultrasonic sensor, camera, and Kinect sensor were all mounted flexibly using an interlocking system. The sensor positions are adjusted so that the quadrotor components do not obscure the FoV. To ensure a clear field of view, the quadrotor is set up in a diamond shape. The vehicle, including its components and payload, weighs around 2,100 grams.

11.3.7 Integration

A system is a collection of subsystems that work together to provide the overarching functionality. System integration entails bringing together multiple, often different systems. It's also about enhancing the system's capabilities, which are made feasible via interactions between subsystems. The proposed UAV is a quadcopter with a

50-square-centimeter footprint. The copter's frame is comprised of carbon fiber and 3D-printed components. Without and with full payloads, the UAV's endurance is 32 minutes and 18 minutes, respectively, having 500 grams of payload capacity. Motor RPM is varied using a controller to maintain the UAV thrust. The controller has an IMU, and is in charge of keeping the UAV characteristics at their best. Using a laptop, the copter's telemetry data is seen and is relayed using a transceiver with 2.4 GHz. The UAV has two cameras, and its computations are conducted on an i5 laptop. The data from the inbuilt camera is sent over WiFi.

Figure 11.9 shows how all hardware is integrated through the software. The RGB-D camera connects through a USB cable with the onboard computer. It can receive and transfer data from the camera to the onboard computer, as well as from the computer to the camera. The optical flow camera connects to the onboard computer in the same way that the optical flow camera does. Through UART (Universal Asynchronous Receiver and Transmission), the onboard computer communicates with the collision avoidance system. The collision avoidance system also receives the data of obstacles from the ultrasonic sensor, and it estimates the travel distance of the quad-rotor and how fair the obstacles are from the quad-rotors path. The FCS (Flight Control System) receives the data from the collision avoidance system and data from the onboard computer. From these both data it decides and gives the command to the control surfaces to move the quadrotor. The overall function will be monitored in the off-board laptop through the WiFi data link.

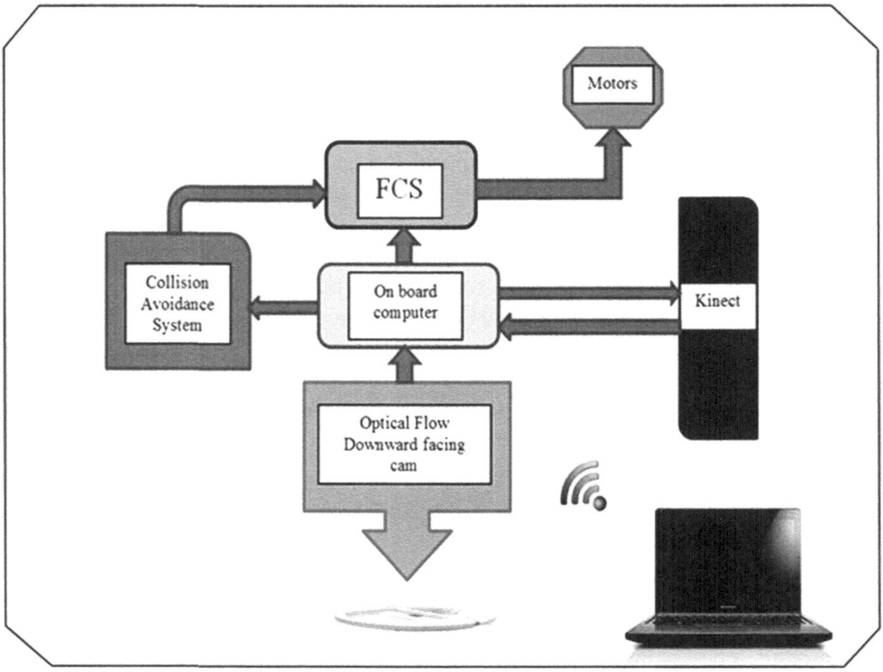

FIGURE 11.9 Overall system integration.

When the vehicle is operated in all six degrees of freedom (DoFs), the quadrotor becomes extremely complex. As a result, the complexity of the dynamics is minimized to three degrees of freedom by maintaining constant height, roll, and pitch for most of the UAV. This forces the vehicle to move in the same 2D plane as it. For picture matching, the SURF algorithm is utilized. The location and x, y coordinates are then determined, as well as an error estimate. After that, the adjusted values are transformed to pulse position modulation (PPM) scheme and sent to FCS. The FCS, which includes gyros and accelerometers, ensures the aircraft's constancy. The optical flow approach of Horn and Schunk was employed.

11.3.8 Brightness Constancy Assumption

The brightness of the image is a major constraint and affects the full optical flow system. The OFN fully depends on the captured images and hence brightness is a major factor to be considered. The brightness constancy assumption is used, which is with an assumption that the brightness is kept constant. The set of equations shows the derivation of the brightness constancy equation. The RGB-D camera suite assisted in creating stable UAV behavior, and the camera delivers attitude control commands to the vehicle; hence, a distributed nodal architecture was created. The use of ultrasonic sensors for obstacle avoidance is an innovative method used on the UAV to effectively complete the mission. The optical flow camera is the onboard downward-facing camera that is used to identify and track the object with pictorial data input. The real-time processing is done off-board and streamed via WiFi.

11.4 RESULTS AND DISCUSSION

The OFN involves multiple steps to accomplish. For each step, results are taken, studied, and analyzed. The results are varied with various inputs and the reliability of the optical flow process is studied. Considering the input characteristics, the results are studied. A reference image was chosen and given as input to the processor. The onboard processor processes the input and searches and matches the obtained image. Thus, the first step of the optical flow process is initiated. Figure 11.10 shows the initial result of the image matching process. This is also tried with various inputs like Lays pocket, I robots, and micro quad.

The second step of the optical flow process proceeds to the step of motion detection. The vector arrows indicate the direction of motion. Figure 11.11 contains arrows pointing to the direction of movement of the substance. The UAV moves over the object based on the direction in which the arrows are pointing. Figure shows the vector direction that shows the movement of a small object. The downward-facing camera placed onboard the UAV is illustrated in Figure 11.12. Figures 11.13 and 11.14 display the trial results of different objects. The next step is brought about after the comparison and analysis works done by the onboard computer. The computer sends PPM signals to the FCS, which controls the movements of the UAV. The UAV faced the problem of maintaining the altitude during its mission, which ultimately led to missing the target.

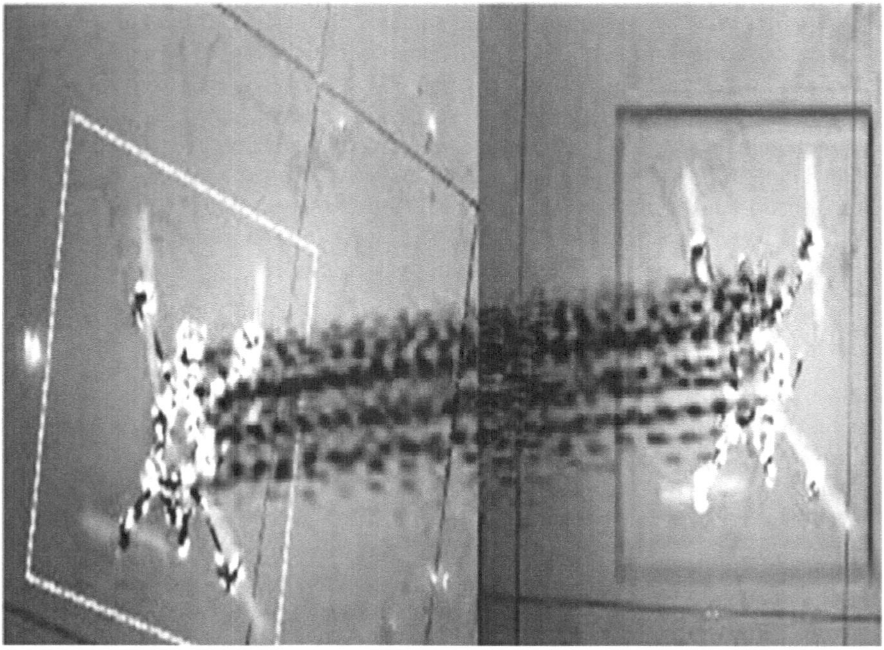

FIGURE 11.10 Image matching process.

FIGURE 11.11 Direction vectors.

FIGURE 11.12 Downward-facing camera.

FIGURE 11.13 Tracking of Lays packet.

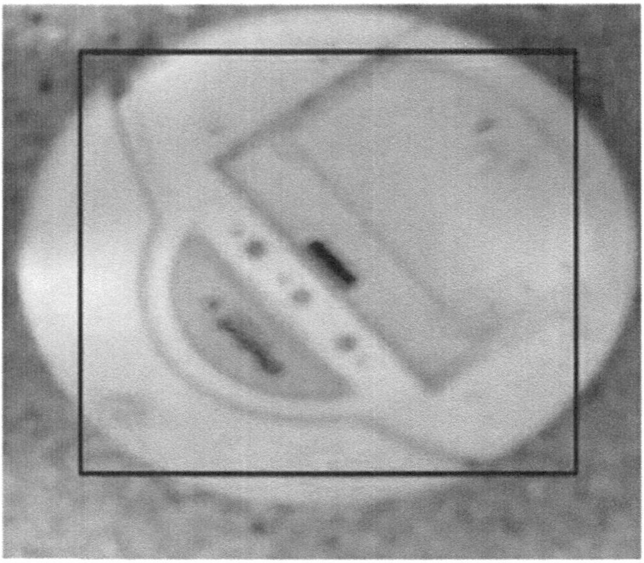

FIGURE 11.14 Tracking of IRobot.

11.5 CONCLUSION AND FUTURE SCOPE

Working on this competition theme helped us to think and develop technologies that can facilitate soldiers to solve real-time problems. Moreover, this competition made us work together as teams and has created an inner urge to serve the nation. The findings of the experiments suggest that using optical flow-based control to maneuver the UAV through indoor corridors is possible. The UAV completes its task in indoor situations by combining OFN with collision-avoidance technology. We hope to test in outside conditions in the future. This form of navigation is extremely useful for the police department, but it may also be used for other purposes, such as guiding a cluster of people from one location to another. This approach is extremely useful in the realm of aviation for rescue operations. The OFN system can potentially be employed in ground vehicles for military purposes.

REFERENCES

[1] S.P. Algur, & S. Venugopal (2021). Classification of Disaster Specific Tweets - A Hybrid Approach, 2021 8th International Conference on Computing for Sustainable Global Development (INDIACom), 774–777.

[2] A.A.R. Alsaeedy, & E.K.P. Chong (2019). Survivor-Centric Network Recovery for Search-and-Rescue Operations, *2019 Resilience Week (RWS)*, 71–77.

[3] K. Jayalath, & S.R. Munasinghe, Drone-based Autonomous Human Identification for Search and Rescue Missions in Real-time, 2021 10th International Conference on Information and Automation for Sustainability (ICIAfS), 518–523.

[4] J. Anand, & T.G.A. Flora (2014). Emergency Traffic Management for Ambulance using Wireless Communication, *IPASJ International Journal of Electronics & Communication (IIJEC)*, 2 (7), 1–4.

[5] S. Jain, U. Agrawal, A. Kumar, A. Agrawal, & G.S. Yadav (2021). Simultaneous Localization and Mapping for Autonomous Robot Navigation, 2021 International Conference on Communication, Control and Information Sciences (ICCISc), 1–5.
[6] K. Sivachandar, & Anand J. (2012). Performance Analysis of ACO-based IP Traceback, *International Journal of Computer Applications (IJCA)*, 59 (1), 1–5.
[7] W. Wu (2021). Design of Small Unmanned Aerial Vehicle Navigation Algorithm Based on Control PID, 2021 IEEE 4th International Conference on Information Systems and Computer-Aided Education (ICISCAE), 514–517.
[8] H. Li, J. Xu, & S. Hou (2021). Optical Flow Enhancement and Effect Research in Action Recognition, 2021 IEEE 13th International Conference on Computer Research and Development (ICCRD), 27–31.
[9] F. Aleotti, M. Poggi, & S. Mattoccia (2021). Learning Optical Flow from Still Images, 2021 IEEE/CVF Conference on Computer Vision and Pattern Recognition (CVPR), 15196–15206.
[10] Y. Huang, B. Zhao, C. Gao, & X. Hu (2021). Learning Optical Flow with R-CNN for Visual Odometry, 2021 IEEE International Conference on Robotics and Automation (ICRA), 14410–14416.
[11] M.J. Lucena, J.M. Fuertes, J.I. Gomez, N.P. de la Blanca, & A. Garrido (2003). Tracking from Optical Flow, *3rd International Symposium on Image and Signal Processing and Analysis, Proceedings of the ISPA*, 2003 (2), 651–655.
[12] J. Zhang, Y. Ding, H. Xu, & Y. Yuan (2019). An Optical Flow-based Moving Objects Detection Algorithm for the UAV, *2019 IEEE 4th International Conference on Computer and Communication Systems (ICCCS)*, 2019, 233–238.
[13] F. Vanegas, K.J. Gaston, J. Roberts, & F. Gonzalez (2019). A Framework for UAV Navigation and Exploration in GPS-Denied Environments, 2019 IEEE Aerospace Conference, 1–6.
[14] Jose A., Nivetha S., Palgunan M., Rupesh B., & Geetha M.P. (2021). SEC-UR-BIKE, *International Journal of Mechanical Engineering*, 6 (3), 2254–2260.
[15] I. Gorovyi, O. Roienko, A. Pitertsev, Y. Chervonyak, & V. Vovk (2017). Framework for Real-time User Positioning in GPS Denied Environments, 2017 Signal Processing Symposium (SPSympo), 1–5.
[16] N. Barnesa, & Z.-Q. Liub (2004). Embodied Categorisation for Vision Guided Mobile Robots, *IEEE International Conference on Robotics and Automation (ICRA2004)*, 37 (2), 299–312.
[17] E.V. Sibia, M. George, & J. Anand (2014). Content-Based Image Retrieval Technique on Texture and Shape Analysis using Wavelet Feature and Clustering Model, *International Journal of Enhanced Research in Science Technology & Engineering, (IJERSTE)*, 3 (8), 224–229.
[18] C. Baker, A. Morris, D. Ferguson, S. Thayer, C. Whittaker, Z. Omohundro, C. Reverte, W. Whittaker, D. H¨ahnel, & S. Thrun (2004). A Campaign in Autonomous Mine Mapping, *2004 IEEE Proceedings on International Conference on Robotics and Automation, ICRA '04*, 2, 2004–2009.
[19] M.V. Srinivasan, & S. Zhang (2000). Visual Navigation in Flying Insects, *International Review of Neurobiology*, 44, 67–92.
[20] A. Jose, K. Sivachandar, & M.M. Yaseen (2013). Contour-based Target Detection in Real-time Videos, *International Journal of Computer Trends and Technology (IJCTT)*, 4 (8), 2615–2618.
[21] D. Marr (1982). *Vision: A computational investigation into the human representation and processing of visual information*. San Francisco, CA: W.H. Freeman.
[22] S. Omari, & G. Ducard. Metric Visual-inertial Navigation System using Single Optical Flow Feature, 2013 European Control Conference (ECC), 1310–1316.

[23] F.F. Khalil, & P. Payeur (2005). Optical Flow Techniques in Biomimetic UAV Vision, *International Workshop on Robotic Sensors: Robotic and Sensor Environments*, 14–19.
[24] J. Rett, & J. Dias (2004). Autonomous Robot Navigation-A study using Optical Flow and Log-polar Image Representation, Proceedings of the Colloquium of Automation, Salzhausen, 1–10.

Index